The One Of Hearts

Michael Mathiesen

Library of Congress Control Number: 2021915288

From the The Rubaiyat

By Omar Khayyam - 1120 A.D.

The moving finger writes; and, having writ,
Moves on: Nor all your piety nor wit
Shall lure it back to cancel half a Line
Nor all your tears wash out a word of it.

Introduction

Several years ago, I was highly entertained watching an episode of the Jerry Seinfeld TV sitcom. You know the one about the Bizarro Universe? It's the one where everyone in Jerry's circle of friends suddenly took on the opposite of their regular personalities. Kramer was no longer brainless and cuckoo. He was erudite and suave. Jerry was no longer debonair and sexy. He became more like Kramer. George, normally neurotic and anxious, became learned, logical and highly personable. Elaine, instead of the steady, confident, amusing one in the group became haggard and silly and scary.

After that show aired, I think everyone in the world started to see similar things going on in their world that mirrored the way that events flipped themselves and people reversed their fate in this first introduction to the 'Bizarro Universe'. It's become kind of a standard way of explaining things that don't make sense in our lives.

But, more recently, I was playing cards with a bunch of friends and noticed, for the first time, that the packs of cards we were using all fresh and new all had the '1's missing. There was no '1' of spades, no '1' of clubs, no '1' of Diamonds and no '1' of hearts. There were no '1's. The actual full decks of all the playing cards in this universe don't contain any '1's, and instead range from '2' to '10' and then there are the royals, and 'Aces' above the Jacks, Queens and Kings holding more value than the rest.

I asked everyone at the table why there were no '1's in any of our decks and the answers were as goofy as they were varied. No one knew, really. So as soon as I got home, I googled it.

For some crazy reason that hasn't been rectified in two hundred years of card-playing, some lesser-known Queen in Europe somewhere declared a law that playing cards would be taxed and taxed so heavily, that people took the '1's out in protest and this became the norm for all future card decks for centuries to come. Something like that. It's vague. I could make no sense out of it. There's also something about stamps in this nonsensical explanation.

Whatever the reason, certainly someone after two hundred years of paying these outrageous 'Card Taxes' would change the rules so that the '1's could be recognized and put back in. But no one did because no one really noticed the difference. Think about that? In a rational universe, not this Bizarro one, this kind of blatant mistake would never happen in the first place, but if it did, it would have been rectified immediately.

There are too many other reasons that make me come to the conclusion that we're living in the Seinfeld 'Bizarro' universe and I reveal most of the biggest ones in this book. But events are not just Bizzare any more. It's actually the 'Bad' universe we're living in - because events have gone way past Bizarre and are becoming extremely dangerous over here, even threatening to wipe us all out.

The good news is that I have figured out where the 'Good Universe' lives, how it can be used to help us in this one the BAD ONE and it's so simple anyone can do it. It's like a good magic trick, no a great magic trick, no it's actually like the greatest of all magic tricks. At first, it mystifies us, but later after we learn how the master magician does it, we all want to perform it and show off our skills to our friends.

BUT - listen carefully - my card-playing friends - because EVERYONE MUST LEARN HOW TO DO IT. We must learn quickly and start performing this trick right now.

TIME IS RUNNING OUT!

Chapter One

- Instant Karma

I'VE GOT BIG NEWS FOR you. Not a single one of us is playing with a full deck of cards! I mean that literally as well as metaphorically. If you look carefully at any deck of cards today, you'll find that the '1's are actually missing. Don't worry, as the title of this book implies, we're going to go into the reasons for this extremely strange occurrence and how to fix the problem. It's bigger than you think.

So, if you pick up any deck of standard playing cards, used for playing Poker, Blackjack, Bridge, Canasta etc., anywhere in this world, you'll find that the cards go from '2' (a deuce) to '10', then to the royals in order of their power - Jacks, (I believe is some kind of a Prince) Queens, Kings and then finally Aces, which comes out of an old word for 'Unity' and therefore have more power than Kings or Queens or Jacks. In some card games, the Ace can also be used as a '1'. But, it is not a '1' because if it was, it couldn't hold more power than the royals.

BUT, YOU WILL FIND no '1's. There is no '1' of Diamonds, no '1' of Spades, no '1' of Clubs and certainly no '1' of Hearts. The inventors of the deck of cards, for some reason, maybe because they didn't know how to count properly, started the deck with the second number in our mathematical system. This is the kind of problem that I look for in life, one that is difficult to explain, so that I can snap it into other observations that are made about the universe to help create a new theory. And so it goes that I have analyzed this complete failure of logic to arrive at the main thesis of this book.

Most of us have noticed that there are no '1's in a full deck of cards. They all start with the deuces (2's). And yet for thousands of years no one has made any moves to correct it, nor do we even talk about it. We've just accepted it as part of our 'Reality'. (And just in case you don't believe me, go and pull out your own deck of cards and look. You'll find there are no '1's.)

WHEN I FIRST NOTICED the missing cards in the deck, I also said nothing, made no comment about it and neither did any of my poker friends. We didn't think it was very noteworthy at all. But lately, I have begun to realize that this is extremely noteworthy because it is a very easy to understand demonstration of how we can take for granted all of the weird things that go on in our version of 'Reality' and just go about our daily lives as if the universe is supposed to be this way. Let me assure you, things are not supposed to be this way.

I have come to realize that this one major flaw, the lack of a '1' of hearts or '1' of any other suit, is an indicator that we're all living in the 'Wrong' universe. It's wrong because there is no rational reason to leave the '1's out of any deck of cards, let alone all of them. Therefore, this universe must be based on irrationalities.

This also leads me to believe that there are two and only two universes because the Creator would not have made only one kind of universe that is basically irrational, 'Bad' and 'Wrong'. No, there has to be another universe to balance out this one. So, there must be a 'Good' universe somewhere very close by which mirrors the 'Bad' universe that we find ourselves inhabiting today. They probably have other names, but for our purposes, we will label them as the 'Good' and the 'Bad' universes. Sometimes you may hear me refer to them as the 'Good', the 'Bad' and the 'Ugly' universes. But, since there can only be two, the 'Ugly' universe is really just another name for the 'Bad' one.

Perhaps even greater evidence lies in the fact that while our planet is spinning rapidly towards a final extinction of all life that has taken billions of years to evolve, including our own form of life, we are mostly engaged in watching baseball, basketball, football, soccer, rugby, tennis, and even cornhole games. This kind of indifference to the major warning sirens going off all around us is simply not happening in the other, more responsible universe, because the other universe is completely rational and must be balancing out this crazy one we live in and that's how I can tell you that it exists.

The reason I believe there are only two parallel universes and not any other number of them, as some theoretical physicists believe, is because of the way that the DNA molecule is shaped.

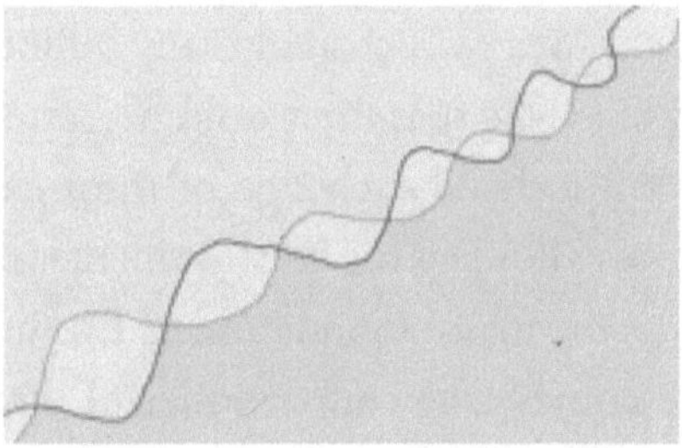

The parallel universes revolve around each other, mimicking the pattern made by the DNA Double Helix.

As we all know today, our fundamental molecule from which we are all constructed has two pathways, two sides of a stairway from which the rungs of the ladder connect them to each other forming what looks very much like a stairway into the Heavens. It's a double Helical shape where the two edges twist around upon each other supporting the rungs of the ladder or stairway.

But in the case of our DNA molecule, the placement of the rungs of the ladder determines every physical attribute of our bodies and minds. The code made from these 'Base Pairs' which are the steps in the stairway varies with each unique individual, but every living thing on this planet possesses this molecule deep inside every cell that makes up that living thing. The code even determines what kind of living thing will result out of the code trapped inside the DNA.

In certain areas of the DNA molecule, the base pairs are telling your cells how to behave like eyes - and they must obey and perform as eyes and you will have eyesight with your own particular eye color and other features strictly according to the code as given to these cells. In other areas of your code of life, your DNA molecule, there are lines of code that determine your height and weight. In other areas, the code sections or snippets, determine how intelligent you might become or even in what kind of occupation you may be best-suited.

We're all dependent from birth to death on our own unique code ordering all of our cells to behave the many varied ways we do. Half of our DNA molecule comes from our mothers and the other half comes from our fathers. This is why 'Sex' was created many billions of years ago. It occurred to the powers-at-be that life would be better off in the long run if there was a well-ordered exchange of information from generation to generation, thus allowing for improvements in the species. This is now known to us as the major foundation of Evolution.

This means half of our fate comes from our wonderful, loving, caring beautiful moms and the other half of our fate comes from our hard-working, reliable, earnest, aspiring Dads. This is all simply basic sexual education we all get in High School - so I hope you're all with me.

But, this human architecture is handed down from Evolution, billions of years of the same molecule inside every cell in your body having gone through countless combinations of the parents of the parents of the parents making up the millions of generations of humans and even other lower life forms throughout millions of years of silent history and then finally landing in your biome, your Sphere of Consciousness (SOC) which you are busy using right now to assess these words as they were written.

In fact, it's well worth pointing out that the molecules that make up your body by the trillions upon trillions hiding inside the nucleus of every one of your cells are the exact same molecules that were once living inside our cave-dwelling predecessors and even the ones that lived inside our earlier ancestors such as the apes, the birds, the dinosaurs, even the insects, and earlier than that, the bacteria and the viruses who were the first expression of this molecule over a billion years ago. This basic formula for life still resides inside of each of us today by way of this code that has been handed down to every single one of us.

We know all this because the code-breakers, our greatest thinkers, the true scientists in this world have been able to hack this code and they can trace the same code patterns all the way back to the opening days and hours of life emerging from the pristine ponds and lakes of our beloved planet.

I know, for many of you, perhaps due to religious texts you have studied, it's hard to follow or accept what I just laid down, but it's simple proven Genetic Science that has been known for centuries ever since Gregor Mendel discovered, in 1865, how much something as simple as peas can teach us.

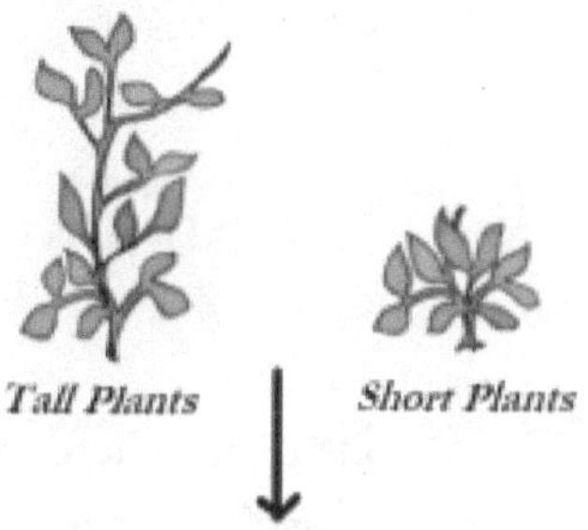

Mendel selected the tallest pea plants and bred them. From their offspring, he could calculate just what the odds were for that trait to be passed down to the next generation of pea plants.

MOST OF US HAVE LEARNED that much of the success of our modern agriculture system owes its existence to the knowledge gained by Mendel about how to grow stronger, taller, pest-resistant or drought-tolerant plants for the growing human population to eat. And, we even get more meat protein out of every acre of land expanded in the same way. We get horses that run faster. We get dogs and cats that we love better. We may even get better humans someday using this ever advancing genetics technology.

By working with pea plants, Mendel was able to work out that half of you comes from your mother and the other half of you comes from your father.

He found: "Each inherited trait is defined by a gene pair. Parental genes are randomly separated to the sex cells so that sex cells contain only one gene of the pair. Offspring therefore inherit one genetic allele from each parent when sex cells unite in fertilization." - Gregor Mendel.

And today we know sex on the molecular level looks like this:

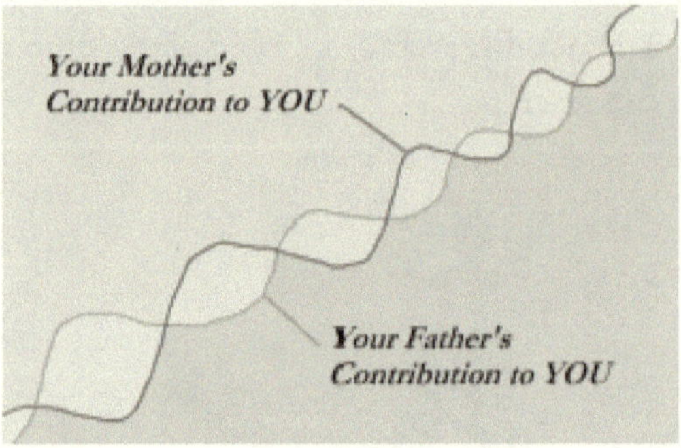

SO, YOU DIDN'T COME out of nowhere. You've been a work in progress for billions of years, as have I and everyone else you can name. Every form of life is a work in progress and every life form is experiencing its fate, the events of living most probable to impact upon it. And even the tiniest of life's events are predicted by the two halves of your DNA molecule, your Dad's half and your Mom's half.

Whatever happened to these two people before you is very likely to happen to you. They came from a different generation, with different places and faces producing different events from your own but taking place minute by minute just like your own events. But, just as the "Devil is in the details", this is where the tracking of each molecular incident that precedes your life gets lost in the fog of time.

Why am I telling you about DNA when we started out talking about playing cards? Because, I needed to show you how the two universes are arranged, and how they must behave first so that you can learn how to recognize the parts where you can make the switchover. I call these points in our lives, the ICOPs, Information Cross-Over Points, the points in our life's events where we can make the switch or are forced to make the transition from one universe to the other.

If you look carefully at the construction of this ladder that is twisting around on itself you will see that there are no real rungs made out of steel or wood or aluminum. These steps on the ladder are made out of another molecule, an atomic structure, just energy flowing across from one side of the ladder to the other. It's not really a ladder. It's an elegantly designed architecture for the optimum amount of energy to flow. The energy flow, just like the energy flowing around in your brain right now, is just 'Information'.

This means there is ample opportunity, millions of opportunities, for your code of life's information to flow from one side of your DNA molecule to the other causing all or most of the variations in your life, the improvements we see as well as the disappointments that weirdly happen and relentlessly take place every day. Genetic scientists call this 'Mutation' or a change in the code, seemingly caused by random events. Cosmic rays, for example, hit your DNA molecules every minute and cause them to change their structure.

But, I believe that nothing is random in this universe and we know this by simply examining past events at this level and watching carefully for patterns.

There's a reason that the code of 'Information' is designed this way. The twisting pathway throughout your life is derived from a larger pattern of 'Information Flow' in the universe and this is all caused by the fact that 'Information' or, as we will see later, 'Consciousness' that holds the information itself and wants this main feature of the universe to be duplicated as many places as possible and as often as possible. Just a theory. I can't prove it. But, this book may hold some of the world's best evidence of this so far.

And, all of this means that your DNA molecule gives you the greatest number of Information Crossover Points from one universe to the other, though in a smaller, more compact model but which we can use and do use often to make our lives far more meaningful, rewarding and purposeful. This is the way I see Evolution. It's much more complex and beautiful than Darwin's correct but superficial conclusions. We don't just survive, we also strive.

I know that I have experienced these cross-over points in my life. I bet that some of you, my readers, have experienced them too. Perhaps you didn't realize what they were exactly and just mistook them for random events that impacted your life for no apparent reason. If so, you probably made that assumption to save yourself a great deal of processing time. Sometimes, thinking about things can be painful, I know.

And what does this have to do with playing cards? Because I believe one tell-tale of the two universes and how you can know which one you're living in depends on whether or not you can go find a deck of cards with a suit of '1's in it or not. I maintain that a universe that counts from '1' is a rational universe and a universe that starts out counting from '2' is an irrational one.

So now you can know that we're all living in the 'Bad' universe - sorry, shocking, but true. In this universe you cannot find a full deck of playing cards with '1's in them, unless you artificially create them or counterfeit them. This is one of the ways in which I have determined that we are living in the 'Bad Universe' (BU) for short. Don't worry, there are many other pieces of evidence for this that I will share, but the deck of cards is the easiest one for most of us to readily accept.

Let's go over some of the other evidence.

In the 'Good' universe we would never have suffered through a World War I, a World War II, the rise of Adolph Hitler and the millions he convinced to hate the Jewish and other kinds of people. We would certainly never have had Richard Nixon as President. Nor would we have had the VietNam War. In a 'Good' universe, it would not be possible to elect someone like Donald Trump or George W. Bush to the most powerful political office in the world - the so-called 'leader of the Free World' since neither of these men showed any talent for leading anybody anywhere good. In a rational universe, this kind of thing just doesn't happen.

We do not see these things in the 'Good Universe' the 'GU' because war is simply not an accepted way of solving problems in that universe. Indeed, there is no Civil War in the United States because there never was any slavery. In the 'GU', when the founding fathers create our form of democracy, they make slavery unconstitutional from Day One. They even create a way for the people to Veto the dumbest ideas that are proposed by the Congress or even the President. This eliminates nearly all of the counter-productive and unnecessary rules and regulations that we are forced to live with in this 'Bad Universe', such as banning the use of Medical Marijuana.

In the 'Good Universe', the Titanic wouldn't have sunk on her maiden voyage due to the incredible incompetence of the people who built it and the arrogance of the man who commanded it. The Titanic, in the 'GU' would have been made of stronger steel and the Captain would have proceeded slower and more carefully around the icebergs that had already been reported to him as in his vicinity.

In the 'Good Universe', there are no atomic bombs laying around waiting to go off because there were no major political threats of this kind. In the 'GU', we would all be fed, housed and clothed and our children educated. Needless to say, with this kind of comprehensive social fabric, Crime is extremely low and there is no need for massive prisons anywhere. Nothing is perfect, so there are always those who choose to make their lives difficult and so this happens over there, but when it does, there's usually a serious attempt to bring these types of folks into the fold.

There are in the 'GU' no abandoned dogs or cats. A tree is rarely chopped down without days of debate among those who appreciate the tree. And, if there's a nest of birds in the tree, it could not be brought down until the young birds fledge and leave the nest.

MOST OF US KNOW THAT it's just not rational or acceptable to the majority to go backwards in Evolution. But, instead of making constant progress from 1865, this country and others around the world, have made major concessions to Evil thought processes and we're doing it again today by reducing and even eliminating our right to vote in this country, the bastion of democracy, the world's last and best hope.

Indeed, today, more than any other time in our million-year history as humans, by making such irrational choices and standing by them over all cogent and the most agreed-upon objections by the thinking, rational people, we are now on the verge of total extinction. The odds increase every day that we will destroy all life on this planet either due to a global thermonuclear war or a slower, more painful end in our constant and uncontrolled production of greenhouse gases that will soon raise the temperature of this planet to over 900 deg. F., hot enough to melt lead, the surface temperature of Venus where no life can exist or will exist forever unless something miraculous happens.

In this universe today, we also have suffered and are still suffering from the Covid-19 Pandemic, a global attack of a virus that has already killed millions of people worldwide and through its variants could end up killing millions more. Worst of all, there is evidence that this virus was fabricated in a Chinese germ warfare facility near Woohan, China where it first began to impact the local population and from there quickly spread around the world.

In the 'Good Universe' severe epidemics may occur, but as soon as they are identified in a few hospitals, the appropriate vaccine is manufactured within days and the spread of the virus or bacteria is kept to a minimum. The viruses don't get a chance to create variants that kill over and over. And, to my knowledge nothing like this virus we are suffering here is ever created by a military lab somewhere to purposely kill our own kind, not because there are any laws against this. This cannot happen only because it's inconceivable to the average human to want to do this. Anyone who even recommends such a thing, would be severely punished in the Good Universe.

Another way you can tell you're in the good universe is that there is a Global Constitution that has been adopted by nearly all countries. Part of that Constitution dictates that lying to another human being is a serious crime, a felony in fact, with severe penalties associated with it depending on how big of a lie was committed. So, there are no conspiracy theorists spreading lies about scientific truth like vaccines, and therefore everyone does the right thing to protect their own friends and family. This single historic document keeps society functioning much more smoothly than over here. People can trust one another and lawyers have a very difficult time making a living.

Needless to say, in the 'Good Universe' there was never a Holocaust or anything even remotely like it because the murder of even one human being for no reason other than their religious beliefs is impossible for most brains to harbor. The German people are able to discuss their feelings in the open without fear of repression and the human brain and soul had developed far enough so that the major emphasis would always be on understanding one another and accepting our differences. 'Viva La Difference', as the French would say - and this kind of spiritual love for one another, as Jesus of Nazareth once described it, rules the world. But, ironically, in the 'GU' no one needs to go to Church every Sunday to relearn these great truths. They simply live in the hearts and minds of nearly every human. Churches don't have too many parishioners over there since there really is no need for them.

And so, it follows that In the 'Good Universe', the religious wars between the Christian and the Moslem world never took place. The word 'Crusades' never took on any political meaning as it does in our universe. There was never an attack on the World Trade Towers, and in fact they still stand there in Manhattan as they were constructed. And, there never was a war in Iraq to take revenge on the terrorists who attacked us on that day because there was never any need for terrorists in this universe at all so they never banded together to kill us.

And, one more thing I will add. In the 'Good Universe' innocent children are never gunned down by deranged people who own enough firepower to take on a country, let alone a school full of our children. People who are mentally ill or overly aggressive as children (bullies) are flagged by family, teachers and friends as being a danger to society and enough of these 'red flag warnings' disqualifies that person to own a gun for the rest of their lives. Luckily, the question of whose constitutional rights are more important, the crazed gunmen or the innocent children has been settled in favor of the children.

Others who may have escaped 'through the cracks' and who commit serious crimes that are determined by the courts to be predecessors to true evil are punished by segregation, not in prisons, but in psychiatric and medical treatment facilities where they must perform socially beneficial work, like sweeping the streets or picking vegetables from the farms until such time that they are they are judged to be safe enough to release back into normal society. But, a second offense or a third such offense results in longer and more socially beneficial work they must perform.

To keep it simple, in the 'Good Universe' things progress along the lines of reason and what will result in the greatest benefit for the largest number of people and other living things on the planet.

In the 'GU' no one would dare harvest a forest so much that it cannot recover. In the 'GU', there are no jet setters able to pollute the atmosphere so egregiously as they do here simply because jet engines were never allowed to be produced. Instead, electric powered airplanes fill the sky because they are silent and powered by the sun, with newer models actually scrubbing the atmosphere of CO2 and giving out oxygen as they travel about the planet, as trees do.

On our side of the universe, by comparison, millions of acres of trees are chopped down by the millions of roaring chain saws, knocked down by bulldozers, burned up in major fires totalling thousands of acres all because of greedy business conglomerates who are willing to kill the planet in order to maintain a rich and opulent lifestyle for themselves and their friends.

Sadly, in the Bad Universe, the minds of our political leaders, due to the worship of Propaganda and lies are full of baloney and conspiracies. There is no room in their brains for critical thinking since all of their neuronal pathways are clogged and muddled with mental 'horse manure'. Sad, but true.

(How many times have you seen this result of a constipated mind? If you really want to live, you must first unconstipate your mind.)

Perhaps worst of all in this 'Bad Universe' there are elected officials who are working to destroy the very democracy that propelled them into office and gave them this much power over us, the greatest betrayal against the people that has ever been known. And unfortunately for all life on the planet, the system they are abusing contains no way to veto, reject or reverse this treason against the people, this attack on common sense, or prevent it from becoming worse.

Total incompetence and outright evil exists almost everywhere you look in this 'Bad Universe'. Somehow we have survived this kind of archaic and 'life-threatening' system of government, but only due to the fact that the problems it created were not catastrophic in nature. Until now.

In the Good Universe, I have to tell you, problems get solved within the first few minutes of their arising and well before they can become fatal problems for the people. I know all of this because I've been in the Good Universe. You probably have as well, though you may not have known you were there and so it may seem like a dream or fantasy of some kind. I'm here to tell you, it's real and this book is intended to go a long way to making it possible for all of us to live more in the 'GU' than in the 'BU'. I hope and pray you will take this challenge upon your shoulders if not for yourselves but for all of the millions of future generations that are sure to come out of, and be forced to live with these choices and who will either cherish us or hate us beyond measure.

In the 'Good Universe' we want you all to enjoy soon, we would not be talking about these horrendous kinds of steady, depressing events. We would not be talking about them simply because they would not be part of our Consciousness.

On the other side of the Golden Helical line of events, we are simply enjoying our lives as God intended them to be enjoyed. Your avatar, your twin and mine are busy working at a job they love and they love their work because everything they do supports not only their family, but also supports a healthy planet. In the 'Good Universe' no one is working to pump toxic fossil fuels out of the ground to be burned by one gas powered, noisy, smelly machine after another from cars to ships and planes, to leaf blowers.

On the other side of the ladder, people all work in occupations that create things and move them around the globe using energy that is produced by sustainable and clean resources such as solar, wind, hydro-electric and even a different kind of nuclear power that we have yet to develop over here because our minds are poisoned by so many of these toxins we are producing that not even our keenest minds can conceive of such a thing here.

And, in the 'Good Universe' I am certain that if you opened up anyone's game drawer and pulled out a deck of cards anywhere in that world, you would find the deck of cards all have '1's in them, the '1' of Diamonds, the '1' of Clubs, the '1' of Spades and the '1' of Hearts.

Why? Because this is simply the more logical and reasonable way to start counting in any kind of game that is based on numbers. We only leave the '1's out of the deck on this side of the Helix because we're all mentally-challenged, but don't even know it.

Part of living in the 'Bad' universe is that we all must live with a major flaw in the development of our brains so that the vast majority of us have a hard time thinking critically and logically. And, this exacerbates the things that go on in this the 'Bad' part of the universe. The 'Bad Universe' exists and if you are living in denial about this fact, I feel sorry for you.

But worse, I feel that the end may be inevitable for all of us if this condition persists and if we don't at least attempt to find a way to burrow over to the other side, the 'Good Universe' to make changes over here that we find over there and we can see are working. It's still not too late for the Earth, in my opinion, if we can learn how to manage the Earth soon. And by 'soon', I mean by 'Yesterday', to keep it simple.

Don't worry, by the end of this book, I believe enough of us will have the knowledge and the powers that are necessary to find the individual cross-over points in our lives so that we can all take the trip to the 'Good Universe', bring back the news of our discoveries and make things 'right' over here, at least 'right' enough to avert the total disaster some of us can see is coming at us at break-neck speed. Keeping fingers crossed.

SURELY, YOU CAN UNDERSTAND the urgency for us to do this - to make a national exploration of the 'Good Universe' study every aspect of it, the laws that they have put in place, the safeguards to prevent any more threats to our continued survival as a species, take all of these superior ways of doing and thinking and then importing them over here to the 'Bad Universe' in time to avoid the biggest catastrophe in history peering from just around the corner. The whole planet is burning up folks, and no one seems to care enough to do anything about it. What more evidence do you need?

This is the half of the universe in which you are all deeply involved just as much as I am. We're in this together and so there is now an experiment taking place in which we must prove to the Creator of all things that we are worthy of all the love, work and effort that went into the making of us. We must not and cannot fail that Creator.

So, think of our current climate crisis as the lab experiment that will decide the whole rest of our future. The chief scientist, the master coder, the chief executive officer, is looking down and watching how seriously or how casually we all take our lives. Are we smart enough to figure out that the way we are doing things is killing us and will we be smart enough to figure out how to counteract those deadly activities and convert our efforts on this planet to efforts that reverse our suicidal march over the cliffs like so many lemmings. This is the one and only question that we must be asking ourselves today and it is the one that I hope to have some real answers to - and specifically designed for your own individual future - by the time you stop turning these pages. But, please don't stop now.

In the rest of this book, I am going to give you every tool you need to make these necessary course corrections in your lives. You must learn them faster and better than anything else you have read or believed or seen before.

And finally, I have to tell you that I am also currently 'assigned' to this 'Bad' irrational universe as you are, so my life is just as tormented by these events over here as yours is. Analyzing them clearly and objectively calculating their origin and their impacts on my life is how I arrived at these theories. With knowledge gained from this brief association of my words we are sharing, you will have gained these same skills if you want them.

Chapter Two

The Crossover Points

IF YOU WILL PLEASE study another view of the structure of the universe for a moment, I will begin to demonstrate more clearly where an ICOP may occur in our lives and what you can do with them.

What we can see of the universe – i.e. everything after The Big Bang is only a small speck in the flow of Big Bangs that have been going on for eternity and will continue happening through the rest of the coding of eternity.

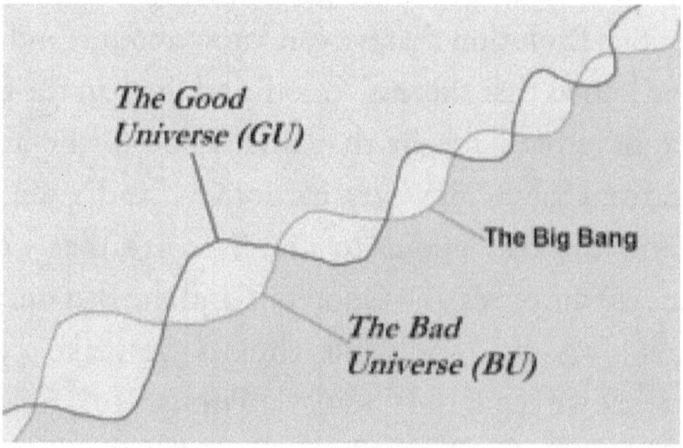

AS YOU CAN SEE IN MY drawing above, everything in our universe covers a small speck in the overall flow of all the universal energy that is ongoing out there. But, even though our known universe is just a small speck in this design, we still have the ability to understand our place in the universe and use our insights to make our lives better by utilizing higher and higher forms of energy. This has been our path through Evolution and I'm hoping it will continue in this manner.

At this point in my book, I hope that you have accepted or at least have an open mind about the way that the universe is constructed. It was an eye opener for me when I first discovered just how amazing the shape of the DNA molecule is and how it formed in this way.

When the universe first formed, there were two paths it could take across Space and Time. The two paths formed a shape similar to the DNA molecule and so it is that the code of all life in the universe mimics the shape of the two universes as they move across Space. This is our greatest point of perspective. This universal model is maintained so that the proper 'Information' can flow from the 'Head' of the universe all the way down into our heads. There's no other way to say it.

I believe this strongly because of the new knowledge we've obtained about how information flows around and through every part of our bodies via our DNA molecule and also how information flows around our observation point here on Earth from the most distant points in Space. And, we're getting this information at precisely the same time in our Evolution that we can know about it and assess it.

Everyone knows that there is 'Good' and 'Bad' in the universe. For centuries we have been taught that this is due to the ancient battle between God and Satan. However, it's not God and Satan that delivers the Good and the Bad events to our lives, it's that we wander in between the two universes, the Good one and the Bad one, at different points in our lives. It's simply the choices we make, based on the information that we get at the Crossover Points.

Attached to that finding is the notion that the copy of the map through the universe serves as a code of life and that every code of life has within it, the probability that part of our lives will be 'Good' and the other part will be 'Bad'. Add to that, that every living creature on this planet has this same molecule inside every cell that makes up that creature - every single one since time began on Earth, so there is no way that any form of life can escape this formula.

In fact, I believe strongly that whenever humans discover life on another planet, every individual form of life on that planet will ALSO share this same molecule because it is a code of life that reflects the design for the universe as a whole. Why would the master programmer of the universe use a different form of coding for every other planet out there where life exists, when this one is designed so perfectly? Hence, the shape of the DNA molecule is literally a universal pattern or template and is the only way that energy flows from the smallest amounts of it in the subatomic world all the way up to the largest things in the Astronomical plane.

For this reason I believe the Master Programmer does use the same coding language for every place in the universe where he or she wants life to form. I could be wrong about this and may be proven wrong when and if we land on Mars, or some other planet and discover new forms of life or the fossils of long dead forms of life that did NOT use our same DNA molecule, and used something else instead. But, I doubt it. I believe that when we do find life on another planet, it will be shaped by the same carbon based DNA molecule that creates our 'Reality' here on the Earth.

With evidence presented here and other evidence that comes from your lives, I can now show that the universe is shaped exactly like the DNA molecule, or more properly stated the other way - that the DNA molecule is shaped exactly like the two universes that twist and wind upon each other in their pathways in Space.

But, the greatest card trick of all time is done with 'Information' as the Master Magician reveals more and more of his grand illusion with more and more information as time goes on. We're going to cover this subject in greater detail later. For now, just think about everything beyond your Consciousness as simply information.

When light bounces off an object in front of you, the beam of light energy is filled with information about that object. The object itself is shaped out of a different batch of information at a subatomic level. All of this adds up to the object you see in front of you. This goes for every sound you hear, every object you touch or that touches you, every smell you smell, every taste you taste, every step you take. Every event in your life is composed of 'Information' regarding that event as it is processed by your central processor, your brain. There is nothing more that needs to be said about 'Reality'. That's where it comes from by day and where it dissolves into at night as you sleep.

I cannot emphasize this enough because it should be generally held by now that our lives do not carry the same exact coding as the information that codes for planets, galaxies, black holes or cows or else we would all be much bigger and much more powerful beings. The information that codes for planets, stars and galaxies comes out of a larger array that is probably shaped like DNA because that is how the two pathways through the universe have been given their roadmap to follow. Since this larger coding array lies outside of our known universe, I can only speculate on that part of my theory, but it makes perfect sense to me. The logic here is overwhelming, is it not?

Remember also that all information comes in waves. Hold that thought.

I HAVE FOUND BY CAREFUL examination of my own life's events, that my life's code is spattered with waves of energy preprogrammed into it so that it might mesh better with the larger forces swirling all around us. Gravity, for example, is the force that attracts the larger bodies in the universe and holds things like our own planet in orbit around the sun so that life may exist. It is also the force that holds our feet to the ground. Therefore, gravity exists in the form of information so that our own form of information may arise. It's not an accident.

And so we all exist within the same gravity field and we all exist within the same information field that creates Gravity as well.

In your own life have you never experienced another person who had a 'gravitational' type of impact on your life? This would be a major love affair, a parent or guardian, a co-worker, a celebrity in the media? I think that most, if not all of us, have had that kind of experience or are even lucky enough to be enjoying one now. This would be a person who influences your life so much that you know your own fate is tied to the fate of your other person.

For this and many other connections with the laws of physics coming up soon, I believe that the evidence is more probable than not that the code of our lives is directly proportional to the code of information that the universe follows over billions of years to achieve whatever it is achieving in its pathways through and twisting around our lives. If this is true, then the universe itself must also be a living being, which I have believed since I was a child. Now, I can present you with the evidence of this belief, which I hope beyond hope will be shared.

Of course, none of this could hold the ring of truth unless you buy into the theory that there are more than one universe existing in the same space at the same time and that everything is intertwined by 'The Code'.

I'm not going to go into detail here about how or why many theoretical nuclear physicists believe this concept to be the best explanation for all the information that we now have about how matter behaves in the universe. All you have to do is google 'the Multiverse Theory' and many pages of scientific data and the latest thought will pop up, some of it with extremely complex mathematics being discussed to back one idea or another. Don't sweat the Math. It's really quite simple as I will explain later.

I think I make the case more strongly than any physicist in the previous chapter about how there is every indication that there must be a parallel universe to our own because of the vast number of unexplained phenomena that we all recognize as 'Bizarro' nearly every day of our lives. The one I want you to remember is that there are no '1's in a deck of cards. There's absolutely no good reason for this. It's just a part of our reality on this side of the Universe, and the one I call 'Bad'. At least we have it as a tell-tale.

For now, let's just assume that these physicists are right, that I am right in noticing the shape of the two parallel universes are mimicking the DNA molecule, or vice-versa and let's go from there.

So, I need you to keep safe in your mind that the shape of the DNA molecule is molded out of the swirling energies that creates these two universes and that the design of the basic pathway for all of the stars, the galaxies, the planets, the asteroids, the comets, the very formidable black holes forms the same exact pathway guideposts, a template, that the events in our lives must take as well.

This leads me to believe that there are crossover points between the two pathways. I also give as examples of this part of my theory that people come in and out of our lives who seem to be from another universe, or at least another planet – am I right?

We can even list some of their names. Franklin Delano Roosevelt, Albert Einstein, Mahatma Gandhi, Buddha, Christ, Lao-Tse as some of the most blatant examples of people who were far beyond the ordinary creatures we know as family and friends.

And how about that unique individual who raised the hair on the back of your neck or sent chills down your spine or gave you goosebumps or took your breath away? How about that one-in-a-million incredible person you've met who you know is your 'soulmate' even though they may not know it or feel it on their end and it eventually breaks your heart?

These times of heart-break in our individual, seemingly unimportant lives are all indications of an ICOP, the Information Crossover Point, in your own life because when it doesn't work out between you, despite all of your hard work and best of intentions and your highest and best behaviors, your most unselfish attention, it could only be because you are each from opposite ends of the parallel universes and the two life paths of these two opposite individuals are never going to be compatible because there just aren't enough Crossover Points for both paths to share a common ground. These two paths will be forever alien to one another.

I want to think that when hearts are broken in this way it is one of the unintended consequences of the Grand Design.

These points in your life can also be as simple as a movie, a book, a painting, a poem, a sculpture, a natural event like a blood red moon, (I've seen one) that can also act as your Information Crossover Point. You can tell because the cosmic event or the man-made work of art will have a growing influence on your thinking process for whatever reason and you can point to that event for the rest of your life as the reason or the pivot point for the change.

As an example, one of my most important Information Crossover Points (ICOPs) was the movie 'Casablanca' with Humphrey Bogart and Ingrid Bergman. I knew as I watched the story unfold, a story they created as the cameras were rolling, by the way, without the usual preparation like screenplays and/or books to follow, that my 'real' life would mimic these fictional events in some strange way even though I was probably fourteen or fifteen years old when I first saw this, the greatest motion picture ever made so far.

Decades later, my feelings about these events and how they would influence my life were confirmed. I had found my 'Ilsa'. I didn't find her and lose her in Paris. I was not the owner of a bar in Casablanca when she came back into my life creating that classic line - "Of all the gin joints in the world, she has to walk into mine." I did not have a piano-playing friend who looked after me. "Play it again, Sam."

But, when in the movie, Ilsa points a gun at Rick's heart and threatens to shoot if he doesn't give her the rare 'visa papers' that she needs to leave Casablanca with her husband, he takes a step closer so that the gunbarrel is inches from his heart, and he says: "Go ahead, shoot, you'd be doing me a favor," I knew this was a major turning point in my life as well because I wanted to feel that much love for another person and I would not and I could not not rest until I did.

Your experience will vary, as they say on TV commercials. You may pick up an extremely well-written book that sings to you, the words jumping off the page, creating scene after scene of a life that depicts your own life in great detail. It could be a classic painting like the Mona Lisa that has this effect on you, as you marvel at the skills of the artist DaVinci whose brush strokes capture the emotions in that famous face that buttress the tedious and superficial events in our daily lives.

It could be a special day at the beach where the waves splash all around you prying open all of your basic instincts, whispering about the millions of years of Evolution just beneath the surface. It could happen out of a very vivid dream or it could come out of a rude awakening.

It could be a shooting star brightly blazing across the night sky just at the moment when you are at your lowest point and so it seems to be screaming in the darkness to give you courage and you get the courage enough so that you are able to pick yourself up and make it through another day.

Your Information Crossover Point could be something like this or something completely different. You know them when you experience them. One of my earliest memories as a child was an event that happens to so many people we have a name for these events. They're known as 'Deja-Vu', which is French for 'Already seen' or Already witnessed'.

I remember sitting in my school bus in elementary school. I looked out through the windshield in front of me and I saw a long line of yellow school buses parked ahead filling up with kids on our way to a field trip. I had the distinct feeling that I had been there on that exact same bus and at that moment before, that I had experienced this particular event in my life before. The feeling was so strong, I never forgot it and I have had a few other events like this in my life, but none of them as vivid as this first one.

I believe that if you experience one of these 'Deja-Vu' moments in your life, it's most likely due to the fact that you are actually experiencing an involuntary leak of information across the boundary between your two universes that I call the Information Crossover Point (ICOP). The normal scientific explanation that this is merely a kind of 'cerebral feedback' in your brain is an easy and vacant one.

Next, I need you to understand what I mean when I use the word 'Information' that can flow across these 'Information Crossover Points'. For decades now, physicists have discovered more and more about the universe we live in by studying the information that slaps them in the face via the most extremely sensitive scientific instruments of all time.

The best example of this that I can show you are results from recent experiments out of the Large Hadron Collider and other so-called 'Atom Smashers'. There are several of these humongous machines in the United States, the one at Fermi Lab being one of the most famous along with the Stanford Linear Accelerator at Stanford university in Northern California.

Courtesy LHC - Cern

THESE MACHINES, THE largest and most expensive scientific instruments in our history do nothing more than smash atomic particles together by sending them flying around their pathways at nearly the speed of light in opposite directions and then bending their pathways through the collider, using the world's most powerful magnets so that they do just that - collide with one another.

If you've ever watched a 'Destruction Derby' a car race where the cars cross over the tracks in the middle in opposite directions as fast as they can and smash into each other head-on, you know how these machines work, just on a much less meaningful level.

In the collision, physicists study the mathematical data, often in the terabytes, of what happened after the collision of any two subatomic particles. They study this data for years, if not decades and glean whatever they can about the subatomic forces that are part of us all. It's not just one or two electrons or protons they study. They study billions of them at a time. All of this information comes to them from the data that is collected about the particle collisions and which takes place in a fraction of a second.

The collision of two subatomic particles at nearly the speed of light

THE POINT I'M MAKING here is that none of these subatomic events that the particle smashers make are actually viewed by the naked eyes of the physicists. We have yet to see something as small as an electron or proton with our own eyes. Yet, the information that these things exist is constantly coming into contact with our brains and the result of that contact is a series of theories about the universe that we call the 'Laws Of Physics'. Out of these laws of ours, the universe takes shape in our minds. These shapes and images are merely secondary forms of information saved as electronic points in our brains.

These laws are constantly being altered and updated by new information that flows to the human brain via our latest scientific experiments and equipment. And so as we progress in our scientific prowess, so the shape of the universe changes as well. It's from this equipment buried deep in Switzerland and France, the largest atom smasher on the planet, that they could announce to the world on July 4th, 2012 the discovery of the 'God Particle'. Don't worry, we'll get to that.

Now, I must not go any further without giving you another example of the 'Information Flow' in the universe that comes to us from another major though recent scientific breakthrough.

Below is a picture of the latest scientific achievement in the installation of the 'Laser Interferometer Gravitational Observatory' or LIGO for short. There are now several of the most sensitive instruments we know placed all around the world. They are about three kilometers long on either side.

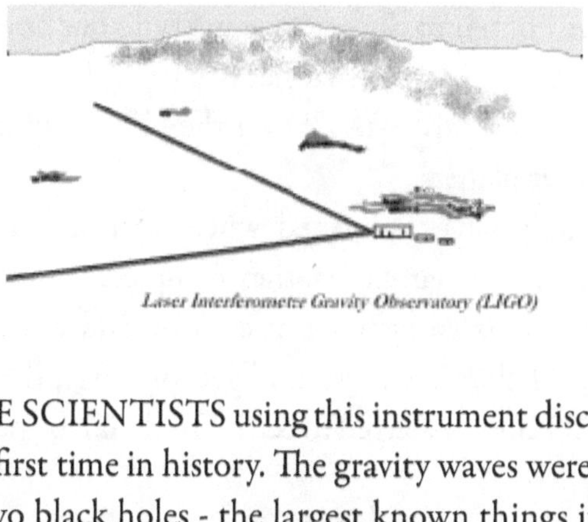

Laser Interferometer Gravity Observatory (LIGO)

IN 2015, THE SCIENTISTS using this instrument discovered gravity waves for the first time in history. The gravity waves were caused by the collision of two black holes - the largest known things in the universe - into one another. The odd thing about this collision is that it took place over 1.3 billion years ago and the news of the collision, the actual gravity waves took that long to hit our detectors, the LIGO equipment here on the Earth. The most incredible part of this story, however, is that they hit the Earth, the same day that humans were first able to turn on this instrumentation that would alert us to them.

In my unconventional way of thinking, it appears to me that someone or some thing caused these two massive objects to collide with one another with the precise timing that we humans would need to evolve here on the Earth and then be able to listen for their collision as part of our 'Wake-up call', some 1.3 billion years later.

If you want to read more about this event in order to verify what I have just said - please do so before you go any further, because this very strange string of events is an important bit of evidence that I use to make up the rest of what I am proposing to you here.

Suffice it to say, the LIGO incident is exactly the same as you or I having the ability to go to the beach - carrying our surfboards - on MARS, where there hasn't been any signs of flowing water in over a billion years. But somehow, we just know that this condition will change somehow, so we construct our surfboards and make our way to an ancient beach on Mars and then, after sitting there for only a few minutes, the entire basin on Mars fills with clear pristine ocean water and so we jump into the ocean, jump onto our boards and at that precise instant, the only wave that has been seen here in over a billion years picks us up and carries us all over the place until it collapses and moves on to other planets.

This is exactly what happened when after spending billions of dollars and billions of man-hours, they completed construction of the LIGO experiment six years ago, turned it on and went home to bed. Before they could close their eyes and get some well-deserved rest, the physicists at LIGO were alerted to return to their lab and observe what had just happened.

They had gathered the 'Information' that the first Gravity Waves ever detected had just flowed over the Earth before sweeping over us, going back out into Space - on the first day of the machine's operation.

This was so unnerving to these scientists that they didn't tell anyone about it for six months, believing that there must have been some kind of glitch in their data. Something had broken, or someone had hacked into their computers and planted fake information, or so they thought. The event was simply too coincidental for them to accept it right away.

After six months of conferring with all the experts in all related fields of Science they finally had to announce to the world the reality of the event and what it meant.

Someone or something caused two black holes to collide 1.3 billion years ago, and 1.3 billion light years away, so that, by the time the news of this event reached us we would have the equipment all set up and ready to receive it. Accidental coincidence? A random event? Of course not. There are no accidents in the universe.

The most profound piece of corroborating evidence that I use to substantiate that this was no accident is that the Gravity waves when turned into sound - composed a short 'WHISTLE'. It was all the proof that I needed. When someone whistles at you - what do you do? You pay the proper amount of attention because your life may depend on it. And so I did.

Don't BELIEVE ME? - The, please click on this link to hear the sound of Gravity Waves colliding in the universe just in time so that we could not only see it, but also HEAR it. Then, make up your own mind.

https://www.ligo.caltech.edu/video/ligo20160615v2

Now, if you have such a cold heart that you can listen to the sound of two black holes colliding - the first to be discovered by humans - and NOT realize that this can only be a signal from the master programmer of all things in the universe - then I am extremely sorry for you.

If you don't understand this, you need to concentrate on the workings of your own heart. You need to learn how to listen to your heart more than your brain and realize also that it is important that I'm available right now to tell you about this and that if we don't all concentrate on listening to our hearts more and more - we're doomed as a civilization and as a species. Because I riddle you this – 'Why else would an event of this kind be coming to us at this time?'

Important enough?

For those who get it - let's continue. I'm going over all of this most important science stuff here because the major overall point to be made in this chapter is that everything in and about the universe is simply 'Informtion', but information that is given to us only when we're ready to absorb it and use it for our benefit and our survival. And with this latest scientific evidence I've just shown you, I posit that our survival means something to the Creator of all of things great and small.

So, what have we learned so far?

In our observations of the smallest of things in the universe, we find evidence that these things are actually only tiny bits of information. They don't really exist at all in the sense of being something solid that we can put our fingers on and taste or smell or touch. The subatomic particles, it is now known, are only tiny waves of energy that makes up the larger fields of energy, not enough to light even the tiniest of light bulbs and yet they are powerful enough to find each other, latch one upon each other in the trillions-times-trillions of combinations until they give all living creatures like us everything we need to exist.

And in our studies of the largest of things, we find evidence that the biggest things in the universe are all just information too. Nothing is solid until we ourselves come into contact with these things. The day we land on Mars, it will instantly transform itself from a bunch of information packets about Mars to an actual place in the universe where we can set our rocket ships on and then get out of them and walk around on the solidity of the red dirt and rocks that compose our nearest neighbor.

And when we use our giant eyes in Space to gaze out upon the furthest distances in our known universe, we are looking at information about the stars and other galaxies, the formation of the universe we live in, that goes back billions of years. When we look up at the night sky, we are actually looking back in time. This means that the present moment of our lives is actually composed of information from the past. It's all happening so fast, the information flows past us so fast that it creates images for our brains that fool us like motion pictures fool us into thinking that the reality we observe is actually the reality of all realities, but it is not. It is merely the illusion of the information flow.

And one more basic tenet of this basic information flow in the universe is that as soon as an event takes place, the event itself becomes more and more just a memory, or a series of images, sounds may be remembered too, but these are also just residing in our memories, unless they are video-taped. When that happens, the images exist in another electronic format, just not in the electronic format in the brain. Therefore, the Past is nothing more than information. And information is nothing more than waves of energy that carry these images all over the place, some of which ends up stored in our brains. And this is how two separate universes can exist in the same Space and Time.

Therefore, when I say that there are 'Information Crossover Points' in our lives, this is what I mean.

Michael Mathiesen's Theory of the Helical Path Universe is as follows:

In the parallel universes that behave just like the DNA molecule coding our lives, there are twists in the helical pathway of our lives through the two universes where we have a chance to help ourselves by observing the coding of information from the other side.

You may want to read my theory over a few times to truly be able to use it.

This means that the book you are now reading could be an ICOP for you. You may have learned enough already to see and feel the Information Crossover Points in your life and are ready to use them wisely. It's truly not Rocket Science. It's actually quite simple.

Let's take stock of how far we've come and summarize what has been learned so far.

One: There are two universes running in parallel to one another. One is called the 'Good' (GU) universe and the other is the 'Bad' universe (BU).

Two: We know that we're in the 'Bad Universe' BU because in all decks of playing cards, somehow, among other things, the 'Ones' have been left out. This is not logical or appropriate. In the 'Good Universe' one built on logic and trust, every pack of cards does start with the '1's.

There are other, even larger indications of this as well such as the political, economic and ecological disasters we're now living with.

Three: The two universes swirl around each other traveling through or creating Space/Time in the form of a double Helix.

Four: In a double Helical pathway there are points of greatest potential for change where information from one universe may cross over into the other one. I call these the Information Crossover Points or ICOPs.

Five: Learning how to use the Information Crossover points in your life may be your greatest and simplest path forward to our success as individuals and life eternal for our species.

So, in the next chapter, I'm going to focus on how to identify and use the Crossover Points in your life to make it better and more rewarding. And, the more you learn how to do this, the more rewarded you will be.

Trust me.

CHAPTER THREE
- Twinning

SO, IN THE LAST CHAPTER, I hope I made it clear how to identify your own true Crossover Points and then, what to do with them, or how to use them in general. In this chapter, we're going to go into the mechanics of doing this in much greater detail because this is a skill that has rarely been used by the general population and yet today we need this skill more than ever. We now begin to change that part of your reality. But you must become better informed about it.

AND, the instrument that will do this all for you without any real effort on your part, without any formal education, without any cost of any kind, without any anxiety or struggle is your heart, or, as we say, listening to it.

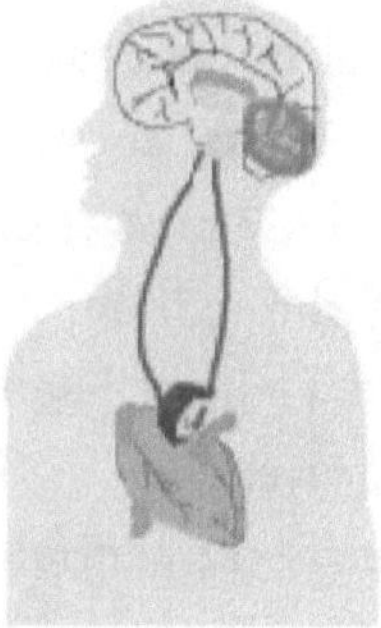

WE'VE ALL HEARD AND practised the art of 'Listening to what's in our heart'. We all learn by an early age that 'The heart knows what it wants'.

When we give the truth to someone from our hearts, that other person understands or should understand exactly what we mean and how we mean for them to take that truth and accept it as gospel.

And, today, Science has found out the truth about these wise old adages. The heart actually houses more than a million brain cells inside it. The heart actually has the capacity to know what it wants indeed and may even be the driving factor in all of our most important thoughts and deeds.

We used to think that only the brain held brain cells, but now we have come to find out that the heart has millions of brain cells too, otherwise known as neurons. What would these brain cells in the heart be doing there if not for a very high level of communication between it and our brains?

It means that information must travel over your nervous system to your brain from the heart and information must also flow the other way from your brain to your heart over this same communication system of neuron to neuron that makes up the network of our brain activity in wave energy.

In other words, your brain is constantly in contact with your heart not just to regulate its beating but to learn how you feel about things, not just about how you visualize, hear, taste, smell and touch them, but how you feel deep inside your core about them. Then, and only then, the brain knows what to do, how to react in your best interests. Having part of our perception flow through our hearts, we can avoid as many knee-jerk reactions, like the 'Fight or Flight' system that land us in so much trouble, or help us survive.

Our brains then, are really the central processing units for all the information that we take in moment by moment in regards to our environment. But, it is the heart that truly feels its way along in the events in your life. And, it is the flow of information between your heart and your brain that determines every step you take in reaction to events going on around you. When you take actions in a 'heart-felt' way, you are greatly improving the quality of your life and you can feel it as it happens like the flower must feel the warmth of the sun as it opens for life's adventure each day.

And, now we know that everything that flows around in the universe is really just information. Therefore, most of the time, we can identify our Crossover Points by listening to our hearts every minute of our lives. Your Information Crossover Points will let you know where and when they are happening through the gut intuition you build up using the wisest, most learned part of your thinking machine.

It's really that simple. I can't guide you to these points. No one can guide you to them. They will reveal themselves to you at the right time for you.

SINCE THE INTRODUCTION of the Internet to the world at the end of the last century, we've all known about and used "The Information SuperHighway". The information about any topic flows all around us at all times through our electronic devices and what we are learning to do more and more every day is how to tap the energy of that information and use it for our own purposes.

But, the Information Superhighway is only our planet's access road to the larger 'Cosmological Information Network' (CIN). The entire universe is made out of information, as we have seen. Therefore, one level of information can easily induce the next highest level. The largest amount of cosmic information came out of the Big Bang a little over thirteen billion years ago. This is now accepted as Scientific Fact. And, it happened in less than one trillionth of a trillionth of the first second of the birth of the universe.

Every bit of information that the universe needs to continue the most powerful and infinite spectacle of Creation that we are finding in the skies using the most advanced telescopes and microscopes in our history was imparted in that tiniest space of time and it has never changed. All of the stars, all of the planets, all of the galaxies were put into motion by all of the laws of physics, most of which we know today, and which are uniform everywhere that we look.

These laws are simply information rules and mathematical certainties that every tiny little bit of energy must obey in its pathway through their existence. And, in my humble opinion, all of the energy in the universe has as its final directive, the creation and sustainment of life in as many places in the universe as possible. What else could all of us be about?

We, the most intelligent form of life that we know about, are the beneficiaries of all of the information that is being transmitted, broadcast constantly and repeatedly in our direction. The information, therefore, can be considered to be the biggest gift we will ever know. We seem to know some of this because often we will call it 'The Gift of Life'.

We are learning more and more about this information because it was always intended for us to learn about it. It's not a government secret or anyone's secret. It's not even hidden. Every motion in the universe can be observed and tracked and plotted until such time that we can derive the rules for this motion. At this point, a new one of our 'Laws of Physics' is put forth by someone we usually call a 'genius'.

But, call them what we may, these are all simply bits of information that are viewed in their final manifestations to reveal their true nature Like in the children's game – 'Hide and go seek' - they want to be found.

Gravity for instance is a mathematical symbol that is used in many mathematical equations.

The symbol of Gravity = 'G'

We learn in Middle School about Sir Isaac Newton and how he discovered and defined the force of Gravity when an apple fell off a tree in his backyard. He reasoned that the apple fell down towards the Earth and not up into the sky because there was a force of some kind that attracted the apple to the Earth.

Thus his now famous mathematical formula of how to calculate the Force of Gravity, a calculation that is so accurate our astronauts use it to wend their way on their speedy trips to the moon and soon to Mars and beyond, or to simply reach the orbit around the Earth and rendezvous with a space station.

$$F_g = \frac{G\,m_1\,m_2}{r^2}$$

F_g is the gravitational force between m_1 and m_2.

G is the gravitational constant equal to $6.67 \times 10^{-11} \frac{\text{m}^3}{\text{kg} \cdot \text{s}^2}$.

m_1 and m_2 are masses

Although this looks daunting, reading Newton's equation for determining the force of gravity between any two points in the universe is relatively simple and straightforward. (Just don't ask me how he came up with this mess of symbols he calls 'The Gravitational Constant'. It gets replaced by Einstein a hundred years later.)

The force of Gravity between two points is directly proportional to the product of the masses of the two points and inversely proportional to the square of the distance between their centers of mass.

This is called an inverse-square law.

For example, if we double the distance between the Earth and the Moon, the attractive force between them would go down (because it is inverse), and it would go down by a factor of four, the inverse of 'Squaring' something.

It also means that the force of attraction between two bodies falls off twice as fast as the increased distance between them. Haven't we all experienced that extra sense of loss when it happens between ourselves and another body where there had been so much gravity once before?

With this formula we can determine the force of attraction between our own bodies walking around on the surface of the Earth as well as the force that will guide our rocket ships to any destination in the universe we may choose.

See, that wasn't so hard, was it? The hard part was the functioning of Isaac Newton's brain at the moment that the mathematical simplicity of this interaction suddenly came to his mind.

I have bothered to go into the details about this, the most famous of all of the laws of physics, as well as the easiest to understand, in order to demonstrate that the information about the interaction of any two bodies in the universe is nothing more than that - information - plain and simple.

It was difficult for us to discover this little bit of information and took centuries before we would develop a brain capable of this great achievement. But this was accomplished relatively recently, as we all know, and it has been proven over and over by our Space Scientists and it has even been improved by another great 'genius' Albert Einstein, early in the last century when he proposed his 'Special Theory of Relativity'.

This theory of Einstein's which we now hold as another universal law of Physics, and the math underlying it, basically created what we now know as 'Space/Time' which can be thought of as the 'Fabric' that holds everything together in space. And according to Einstein, this fabric exists in all directions and connects all things to all other things in the universe. It's this unseen, untouchable kind of 'fabric' that creates gravity. So, Isaac Newton is important because he was able to prove that Gravity existed and that it could be calculated precisely down to the pounds and ounces of force. And, Einstein is equally important by showing how this great force is generated.

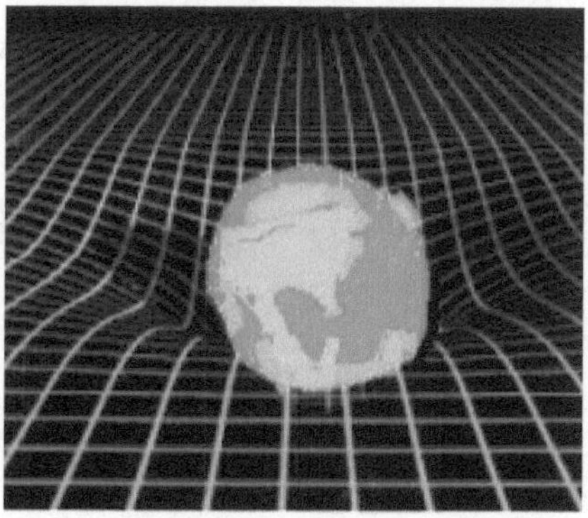

Space/Time bending around a gravitational object.

ALL OF THIS THINKING is now a central part of our laws of physics and laws are merely patterns of events that obeys them by being part of the information flow.

A planet, in other words, is no more or less than the sum of all of the particles that we know can compose all of the atoms and the atoms compose all of the molecules in the universe and all of these can compose a planet.

Therefore, the information about particle existence all forms up to eventually come into existence as planets or stars or even galaxies.

Everything that we see and hear and study and learn about - absolutely everything - is all information. But, oddly enough Einstein's theory about the larger universe all around us doesn't seem to work in the subatomic universe where we find no evidence of the fabric of Space and Time, therefore, none of it "warping", nor do any of these smallest of energies seem to be creating the fabric of Space and Time.

What is described in the larger sense of the universe by Einstein needs one little adjustment that I am calling Mathiesen's Cosmological Constant.

And it goes thusly:

From our perspective inside the universe we can see the how the fabric of Space/Time is made and how it is warped by large

gravitational objects through their gravity fields. But, if we increase our perspective, our point of view, to go out beyond the edges of the known universe, can see that Space/Time is not only warped by larger objects, but is also twisted around itself in the shape of the Double Helix brought about by the forces and fields held within the smallest of objects it contains.

The DNA molecule will also mimic the larger shape of two universes twisting around upon each other making for the great balancing act between all things we consider to be good or evil.

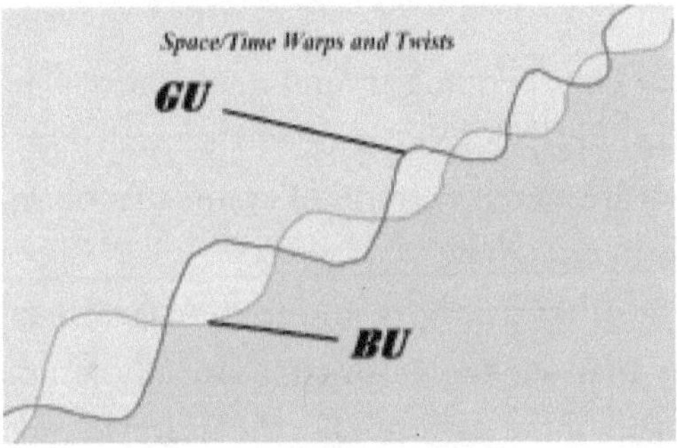

IF WE CAN IMAGINE THE fabric of Space/Time warping around a strong gravitational object like the Earth or the Sun, we can also imagine the same fabric twisting in its course through space time in an even larger and more 'Cosmic' dimension.

Everything that we can see with our most powerful telescopes, everything that goes back to the Big Bang is vast indeed, but it is only one string in the twisted pair of strings that our universe and the opposite universe take in revolving around each other.

But wait - there's more. Everything that happens, every event, is instantly dissolved into the Past and the Past, no matter how near or far, can all be picked up by the billions of brains on this planet which will observe the Past as recorded words, images, sounds, that get converted by the brain again into tiny electronic and subatomic memories only. The events themselves have no real life, only energies that ebb and flow across the fabric of your own Consciousness, something we now know is connected to every other Consciousness out there, big or small by our DNA molecule. And, in all of these conversions, some truth of this information can be lost, depending on the users of these brains and how they filter it.

It may take you a while to grasp this, as it did me. If you don't see this and can actually 'Grok' it in your mind right now, feeling the truth of it in your heart at this point, take some time, go back and reread this entire section until you can realize it fully because the next section depends on your acceptance of this concept, or at least your open mind.

The Information Crossover Point

SO, IF YOU'VE FOLLOWED this line of thinking so far, you are now realizing something truly astounding about our universe and the role that our individual lives play in the grandest scheme of things.

Let us summarize:

The universe has two parallel pathways through Space/Time. This is called the 'Multi-Verse' theory or what I call the 'Two Universe Theory in that there is a 'Good Universe' (GU) and the 'Bad Universe' (BU) that counterbalance each other.

The flow of the two universes in this manner is the manufacturer's mold for our DNA molecule in a double helical pathway that provides a close encounter of the two universes at regular intervals.

And further, that these closest of encounters between the two universes can be seen as 'Information Crossover Points' (ICOPs)

Finally, these cross over points can be part of our own lives and therefore can become known to us and we can therefore take our own fate in our hands at these key points and perform to our highest abilities and even change our predetermined destiny for good or bad. The information can go both ways.

So, we have in front of us two very well-known principles of Science, one showing that we can know and actually calculate the forces in the universe to the 'nth' degree. Indeed, this has been done with all of the other known forces that hold the most elementary particles together.

And, we have another description of the universe, also proven in many ways today, that the universe flows by way of a 'Fabric', actually a four dimensional field of all the known dimensions in Space and adding the dimension of Time which is also only Information.

And now today, in this book, I am asserting that this fabric of Space/Time is actually shaped like our DNA Molecule because this would easily allow for information to flow at certain 'times' or in known and predictible places in the dimension between one universe and the other, while maintaining a perfect balance between all things 'Good' and all things 'Evil'.

So, once again, I would invite any of my readers to examine their lives at this point in your reading and attempt to identify the 'ICOPS' in your own life-paths. They are there glaring brightly in the dark for you to see and know. Some of you may be too young to have any prominent 'ICOPS' and others of you may be old enough to have encountered more than one, perhaps several.

These points in your life can be identified by a day or several days in your life when things seem to go 'off the track'. Results of your planning may go into the opposite of what you had planned for. You may receive something out of the blue that was extremely beneficial to you, but also unexpected.

John Lennon said in a song that "Life is what happens when you're busy making other plans."

You may find a relationship turns sour without warning. It could be a death in the family or a diagnosis of a disease that is truly life-changing for you or a child or a parent. It could be a grade in a class that keeps you from going to college. It could be a person who comes into your life unexpectedly and changes your life dramatically.

For whatever the reason, when a crossover point hits your life's path in any of these examples or in similar fashion, it's your responsibility to identify them and react accordingly.

If you go through your life not noticing and not having some back up plans to these potentials, you are merely drifting from one universe to the other without knowing much about either one and allowing these forces to be your master. The potential exists for every one of us to master these forces and become the greatest form of life that we can be, making the right choices at just the right moments that enhances the reputation, the Grade Point Average, (GPA) of our species, thus bringing us closer and closer to the Creator and what he or she has in store for us all.

I've already drawn you a picture that demonstrates what I'm talking about and so below is my rendering once again of the actual shape of Space/Time.

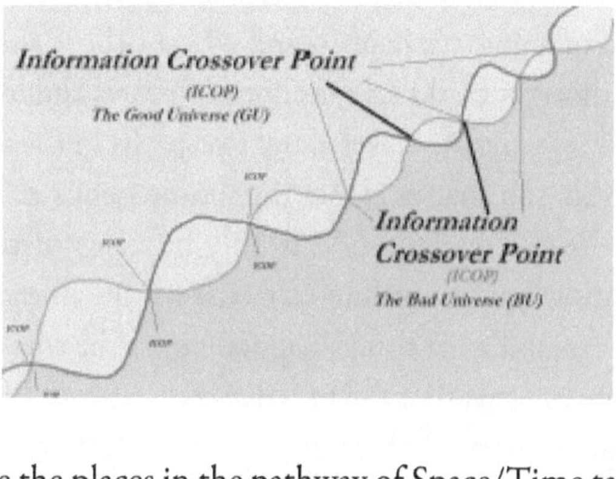

Here you see the places in the pathway of Space/Time takes wherein there are opportunities for an exchange of information from one universe to the other.

THESE INFORMATION CROSSOVER Points can happen at different intervals for every life who cares to observe them. Some of these Information exchanges may be very simple, like learning to ride a bike, or they could be extremely difficult like finding out that someone you love is not in love with you. Some of these periods in your life can be extremely traumatizing, even life threatening, while others can be completely joyous, like falling in love and finding your soul-mate.

Finding one's soul mate, by the way, is how I learned about this shape of the universe because the discovery of my soul mate definitely changed my pathway through this highly energized life of mine, putting me in the 'GU' for a short while. Then, losing her suddenly was the most radical change in my life's path, ripping me out of the 'GU' and back into the 'BU' without any possibility of my reversing it or slowing it down.

I was a very young and foolish person in those days, and so I believe that if I were armed with the knowledge I am now revealing carefully in this book, I may have been able to keep us both on the 'GU' side of the universe for the rest of our lives. But this knowledge was not in my possession at the time.

Fortunately, after losing her, I vowed to change that and dedicated the rest of my life to this kind of research into the heart of things - how this kind of thing can happen and why. I believe the results of my research and thought could save millions of hearts and lives.

So, how does your life travel along two paths in the universe at the same time? Well, the answer is that the simultaneous existence of your life through the BU and the GU is only in terms of the probability of things existing. Since nothing can exist in this intermediate world without the probability of things happening on the smallest of worlds, then, it is probably true that the two instances of your life are traveling along in tandem but at slightly different intervals, if at all, and only under the laws of probability.

When we peer down at the tiniest particles in the universe, it turns out that we cannot observe them and know their trajectory through Space/Time at the same time. We can either know about their trajectory through space or we can know where they are at any given moment. We cannot know both of these properties of matter. This means that everything at this level of the universe is merely a probability that an electron, a proton, a neutron exists where it is supposed to be.

And, this is how I think about my twin in the other universe. It's probable that I exist over there at times and at other times, there is little probability that I exist over there, but the Information Crossover Points always exist for either of us because there has to be a balance maintained between the overall probability of both universes to exist, Good and Bad.

And so, everything we know about the workings of the mechanics of the universe is that it's all a field of energy where we can pinpoint the basic building blocks of the universe such as electrons, protons, muons, gluons, quarks etc., but we cannot know their ultimate destiny.

And, so it must be with our own matter for we are constructed entirely out of these same particles. Where they have no predetermined fate or destiny, neither do we. We can know where we are in the present, but we cannot know with absolute certainty where we will be in the future. We can only guess at the probability of either of our lives' events.

Layer on top of this, the theoretical knowledge that we are more than just the sum or our parts. Humans, and most other creatures with a brain have been granted over time a kind of super-life. Call it Consciousness or the Spirit or the Soul. We all have a layer that we sculpt, build upon, recalibrate, sustain, befriend over time. And this spiritual, persistent vista of our personality array is the only thing that matters at the end of the day.

This energy that is our true construction can flow from one universe to the other at the Information Crossover Points (ICOPS). I admit here and now that this segment of my theory is hard to prove, so I must leave the final judgment of this section of the book to you, my reader.

I know it is true from my own perspective because I have been on either side of the universe - the Good and the Bad. I have experienced what it's like to live in the 'GU' and I am experiencing every day presently what it's like to live in the 'BU' as we all are. And again, my latest and final clue was the lack of a '1' of hearts in all our playing cards. That's what cinched it for me, as I have said.

Until you have made a crossover yourself, you may find it hard to accept that this is possible. As more and more of you make the journey, I believe this knowledge will become mainstream thought and we will all be able to get things done much more smoothly and cooperatively after that process takes shape in our society. But we must get started soon for this side of the universal flow - the 'Bad Universe' looks to many of us as if it's quickly coming to an end.

For now, while we're in this half of the universe, think about how much happier you could be if you were living in the 'Good Universe' and how much better off we'd all be plus all other living things on this planet. After you have a firm grasp of this idea, think about how you might get there. What would you have to do differently to make the jump to the other side? Do you even want to? Or are you happy here where everything is 'cattywampus' and 'Bizzaro'?

Now, remember, every person, place or thing in this universe, on both sides exists only as a probable series of events which, compounding on themselves, creates that person, place or thing.

This means that your counterpart can exist in the 'Good' universe and you can see this happening if you increase the probability of that existence on the other side by merely examining your life in greater detail than you have before.

How do you do that?

First, consider the simplest laws of probability. Flip a coin a million times and over time, the coin will end up as a 'Heads' in roughly 50% of the times you flip it and 'Tails' the other 50% of the times you flip it. This is true because there are only two sides to any coin and it must land on one side or the other. There are no other possibilities.

With human behavior, we know there could be a multiple of possible outcomes from any one of our decisions, and therefore calculating the probability of the events occurring in our lives is more difficult than predicting the flip of a coin. Not even the fastest computer can do it with any real-life accuracy.

My personal belief is, however, that the more familiar you become with your own life's events, their causation and their resolution, the more adept you will become in knowing the probable outcomes of events on both sides of the universe from now on and into the distant future. Thus, in a way, you can live on both sides at the same time. However, one of our life paths will be 'Real' and other one 'Imaginary', or an artificial probability that can be part of the field in that universe, but not a known reality. In effect, you should attempt to spiritually 'Twin' one side or the other - if you choose to.

You accomplish this by simply imagining what your 'Twin' on the other Universe would do at this juncture in your life. This is much like when Albert Einstein used his 'Thought Experiments' to discover our current version of Gravity and other forces.

Like Einstein's thought experiments, your 'Twinning' should be centered around looking at your situation from the perspective of life in the other universe, the 'Good Universe'. How would things be progressing over there in relation to your life over here?

The worst thing that could happen is that you will either fail to learn anything significant about yourself, or you may learn how to take advantage of the crossover point and actually make your reality move over to the 'Good Universe' and actually make a clean transition to it physically. As I have said, this happened to me once when I found my soulmate. I was actually there physically in the 'GU'. And so I can tell you that it is far from impossible, just improbable for most of us - until now.

And, I must interject something here to help you in your own evaluation of your life. The knowledge I have gained about how the Creator of the universe gives all living things special connotation comes from earlier research of mine which I produced in a book entitled:

The 4 States of Consciousness.

IN THIS BOOK, I EXPLAIN how consciousness is formed and how the consciousness of living things is induced out of the existence of 'non-living' things. However, it is also shown how there is no such thing as a 'non-living' thing.

And so, I will summarize the findings in that earlier work of mine thusly. (A quick little online course can be found here - Udemy.com/4statesofconsciousness[1])

Electrons have been found recently to sustain a kind of Consciousness. This is my own conclusion after looking at the data surrounding the phenomenon known as 'Entanglements'.

When two electrons are forced to share the same space within an atom, they become 'Entangled', which we can observe because when we separate these two lovers, they remain aware of each other's position

1. http://www.udemy.com/4statesofconsciousness

in space and also their state or condition. We know this because when we change the 'Spin' of one of the separated 'Entangled' electrons, the other one changes its spin instantaneously in recognition of the change in their partner, spouse, lover, as I like to call them.

This is also evidence for how we humans all have the basic instinct to find our own soul-mate. The quest to find our person starts at the subatomic level of the universe where all things try to do the same thing. We're born with this desire because we're made out of stuff that needs to pair up also with a sympathetic being.

When we find that person, that singular person who completes us and we can't forget her smell, her gracefulness, her warmth, her generosity of spirit, her softness, her voice, her touch. And when you know that the last thought you will have when you pass from this place will be about her or him, you have reached your highest state of existence. There is no other one higher.

So, in direct response to something that affects the electron's partner, they always try to reach a higher quantum state, or at least a different one. How can they know about the state of the other? My answer is that electrons and probably all other subatomic particles contain the property of Consciousness at its most fundamental level.

Since every known molecule in the universe is constructed out of atoms, themselves just a collection of electrons and protons, neutrons, etc. we can assume that this form of Consciousness is induced into the atomic structures in this universe as well. We see that evidence in the fact that all atoms in this universe radiate a property known as 'Rotational' energy. This energy is there so that the molecule can also reach a higher Quantum State. How do they know to do this? The property of Consciousness has apparently been passed up the chain. That's how.

The same thing happens when the molecule radiates the 'Vibrational' form of energy fluctuations and again this happens to allow the molecule to reach another still higher Quantum State. Again, how do molecular structures know that they should behave in this manner? Consciousness again has been passed up to them by the atoms, having it passed up to them by the electrons and protons and quarks that create them.

Since we are all made out of molecules, the DNA molecule having already been discussed, something that orders all of the rest of our molecules how to behave in support of our biological self, the Consciousness that began in the subatomic layer in the universe is passed up to the layer that is our molecular one by way of the Rotational and then the Vibrational energies deep inside us so that all molecules can achieve their highest quantum state. This is intended so that we, as the great collections of these things, can also do the same. Too many of us don't seem to care about this aspect of their lives, it's sad to say. Or maybe they just don't know how.

If you don't understand this aspect of human biology, please google it. All of the movements you take, the things you do every day and think about are performed by billions of molecules in your possession that are told how to behave by our DNA code of life.

But, the highest quantum state that any electron, atom or molecule can achieve is the quantum state that is only achieved by being part of us because our Consciousness can travel the greatest distances in Space/Time as we have seen and are seeing more and more today.

But, what are we best at doing with our form of Consciousness?

We observe the universe around us and give feedback to the original Creator of all the energy contained therein. When we look at the universe and are delighted and amazed, we are giving feedback to the creator of all these things. We are the eyes and ears of the Creator who wishes nothing more than to view this universe through the eyes of a grateful soul. It is my firm belief that when we are the most grateful for our lives and express it, we achieve the highest of all quantum states that all of our personal atoms and molecules are striving for.

Therefore, I have given the name to the 4th and highest state of Consciousness as 'The Feedback Loop'. That's us. That's what we were designed for. We are the ultimate goal of the universe's infinite energies. We can take great pride in that, if we want to.

The best evidence I can give you for all this is that whenever you look up at the night sky and you wonder at the majesty of it all - you are giving the best feedback that can take place in this universe, both sides.

Whenever you rescue an animal that is hurt or abandoned, whenever you plant a tree, adopt an orphaned child, contribute to saving our planet, you are giving the best feedback to the Creator that you know this secret of your own life and the Creator responds to us in powers equal to or greater than our own.

Whenever you resonate deep within your heart and soul with a good song, a movie, a novel, a photograph, a dance or a painting, you are giving the best feedback to the Creator of life that can take place in this universe and you become one with the power and force of Creativity.

It's our job - our greatest purpose in life to give feedback, show our appreciation, to be grateful for all the beauty and goodness that comes into our lives and this is true in whichever universe you inhabit because the great force of Creativity is the way that things came into being the way they did for you and everyone else.

And now you can test this theory of mine. Go and give thanks for your life. Show your gratitude to somebody for anything that you like and watch the reaction and how things move in the right direction, toward the 'Good' and putting all of the 'Bad' in the rear-view mirror.

In other words, by showing our knowledge and gratitude for the derivation of our own lives, in doing our best to reach a higher human Quantum State, we gain ever higher states of Consciousness simply by understanding and nurturing, then improving upon these kinds of events. In effect 'Twinning' the events that could, in all probability, be happening on the other side.

It's not easy, but it's not so difficult that anyone is prevented from doing it.

First, remember that it's only information that you're seeking. I like to think of it as getting to the 'Truth' of your situation. Go back in your mind to the time that you believe changed the direction of your life for good or bad and relive each minute of those days, weeks, months or every moment that you can. And, by the way, it will never be a complete record.

Think clearly about what was said, what events followed the words and how they made you feel inside. Next, try to determine what you could have said or done differently to improve the situation or alter the events in your favor or in favor of your friends and family, if the event was not about yourself.

Once you have thrown around a few different scenarios, try to predict the different future you would have enjoyed in your life, had any of these scenarios taken hold and give each series of events a score in terms of their probability of happening.

Now, here's the good part, the part where you may even be able to make the transition - When you believe you have reached as near to one hundred percent (100%) probability of changing your life as you can, give it a shot. Try to physically make the transition. Take yourself out of the 'Bad Universe' (BU) and get your ass over to the 'Good Universe' (GU). You can do it. I've been there and so I can be your guide. Just pick up any deck of playing cards and see if they contain any '1's in them. If they do, you've made the transition to the more rational, humane side.

Whenever you find a deck of cards that contains the '1's, you know you've crossed over. But, don't reveal the secret, because people in the 'GU' may think you're crazy for even suggesting that the '1' of hearts is left out over here. In the 'Good Universe', they take it for granted to start with the '1's just as we all take it for granted over here in the (BU) that the cards should start with the '2's - the Deuces.

Remember also that it's only the 'Information' that can travel through the crossover, not the physical you. Therefore, simply get more information to the other side that you find necessary to bring the other side over to you.

This may be hard to accept right now, but you cannot physically go there, but you can bring the other side over here to accomplish the same results as if you had gone there.

Don't try to change yourself to fit into the opposite universe. Just get the 'Good Universe' to twist in on itself back to you wherever you are now. In this way, you can flip the two realities to reverse into what we all know is the only way to save ourselves in the long run. It's like taking a reversible jacket, pulling it 'inside out' so that the lining then becomes the outer layer. You may not be able to make the transition physically to your alternative life path on the parallel universe, but you may be able to turn both life paths inside out so that you can start to travel along your life paths that you would choose to take on the 'GU'. The end results are the same.

Of course, this accomplishment is a tall order if we try to do it alone. It could work out much better and happen a great deal faster if we all agreed upon some of these basic principles of making the two universes flip. Therefore, my best advice sitting here today is that we don't try this as individuals. We should all get our acts together and attempt to do it as a united planet.

Think about it this way. If we attempt to reach the 'Good Universe' only as individuals, that might work out OK for the few of us that make the transition work for them, but what about the rest of us? As the Marines say - We don't leave any man behind nor any woman or any child.

IF, on the other hand, we attempt to make the two universes flip as an entire united species, the most intelligent life form on the Earth in order to save the planet - we stand a much better chance of making the flip impact the rest of our lives much longer, deeper and wider.

CHAPTER FOUR

In The 'GU'

WHEN SCIENCE FIRST discovered Anti-Matter in 1932, they began to wonder why there isn't more of it in this universe that is mostly made out of 'matter', the stuff you find yourself mucking about in every day. And, then they wondered if there could be another universe out there where Anti-Matter is the most prevalent type of stuff. And, of course that leads to the bigger question I'm trying to answer here: if there is another universe out there that is composed of the opposite kind of stuff, where would it be? What would it have to do with ours?

In case you are not familiar with Anti-Matter, I will give you a quick tutorial on its major features. When you put an atom together in this universe, you have to have protons, at the nucleus of the atom, and these are always positively charged electrically. So, you'll always see protons denoted with a + (plus) sign.

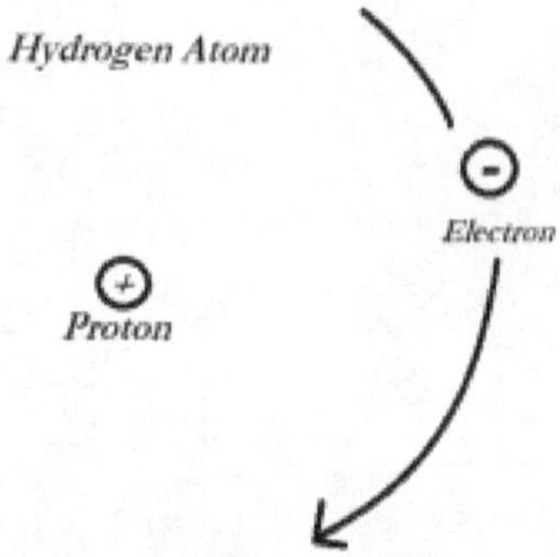

THE NUMBER OF PROTONS that congregate in an atom determine the kind of properties that this atom will display. Hydrogen, the most common of all the atoms in the universe, has only one proton in its nucleus. We call this the atom's 'Atomic Number' and so with Hydrogen, the Atomic Number, or the number of protons in the guey center of it is always '1'. The number of Electrons that will enter this space to balance out the 'Positive Energy' of the proton will also be '1' in the case of Hydrogen. Because there is only one proton and one electron in this particular arrangement of matter, this element will behave over and over countless times in the universe exactly as Hydrogen should which is basically to make the billions upon billions of stars we see in the sky, along with just about everything else.

And 'one' is a special number in the universe because it is with this color of energy that the master artist of the universe uses to paint the most intricate and profound painting of all time, the canvas that we see every night or day, using our telescopes, to see as far back as the beginning of Time. The painting is so overwhelming, so intrinsic to our lives, it inspires all of us in so many subtle and interesting ways, with so much information, it's impossible to know it all.

A necklace of stars mainly Hydrogen Fusion
- courtesy: NASA

WITH THIS TINIEST BIT of information - the 'Oneness' painted all of Space/Time with enough of the simplest form of energy to create all of the stars and planets in the universe. Without this 'Oneness', we get no stars, no heat, no light and life will never exist.

A SIDE NOTE:

And, one additional comment while we're talking about all of the ingredients that make up our universe. When we say that an electron comes in to the atom to balance out the proton, that's exactly what happens. Where there is a free proton in the universe created, an electron with an opposing force will enter that space out of nowhere also and the electron is simply born out of the need for these two forces, the positive and the negative, to balance out each other. No one knows where the proton energy comes from nor where the waves of electron energy comes from either. They simply come into existence from a place that Science has no evidence of.

I think it's easy for some of us to think that there's a proton machine somewhere that makes up all the protons in our lives or there's an electron machine that shoots out billions and billions of them to produce our electricity and other things that we see. But, there are no such machines. These most important subatomic energies are simply created out of nothing. It's the greatest magic trick ever performed.

END SIDE NOTE

It takes a whole bunch of time to cook the universe in this way, so far at least 13.7 billion years or so, but as the old saying goes 'Slow and steady wins the race'.

In other words, the number 'ONE' is the number that is used most often in simmering the stew that is the universe we are all swimming in together.

Another interesting fact about the number 'ONE' is that whenever you take any other number, no matter how large or how small and multiply it by the power of ZERO, that number equals 'ONE'. All numbers, no matter how large or how small are all reduced to the 'ONENESS' by the factor of zero which is not a number, but the lack of number.

So, ZERO (0) is also an important entity to remember because it is an opposing force to numbers that the master Creator uses to paint our universe and keep it in good shape. In a previous work of mine, I show how this fact proves the existence of God, by the way, and so for further information, you should look up that book - 'The Science of Physics - Proof That God Exists' if my logic here appeals.

We also learn at some point about how to become 'One with the Universe'. Or how to become 'One with our problem' whatever it is in order to solve it. So, in order to truly know something you have to become 'one' with it. Easier said than done?

So, 'One' is not only the first number that is created in the universe by the Master Creator of all things, it's also the end result of two or more people or things joining together in order to 'know' something in its highest form. Hold that thought.

But, when we look at the next atom, Helium, it will always have two protons in its gooey center (we now know there are things called Quarks that make up the gooey protons) and two electrons floating around in orbitals with negative charge to counterbalance the protons with positive charge.

And so it goes all the way up the Periodic Table of the Elements (Remember that chart in High School?) as we create more elements by simply adding one more proton, which creates the need for the opposing Electron to be added as well and so on and so forth through a hundred and eighteen total elements that make up our known universe.

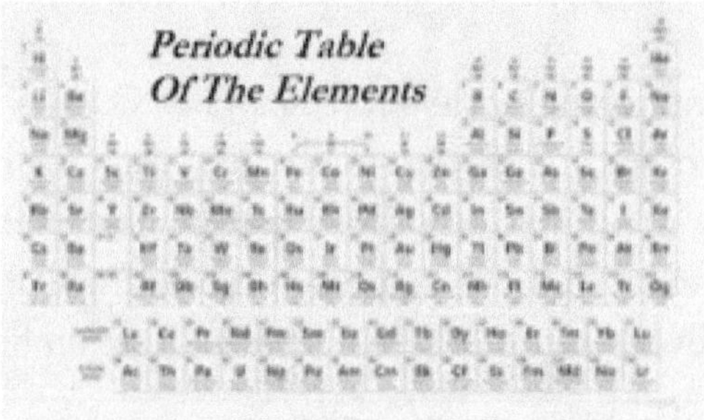

WHEN WE TALK ABOUT these different varieties of matter, we are basically talking about the composition of all things, that which makes up all material things in the universe. And so when we encounter Hydrogen or Oxygen, or Carbon or Silver or Gold atoms, we are dealing with atoms that all have positively charged protons and an equal number of negatively charged electrons to balance them out. This balance of forces is how our universe is constructed, or so we thought for a long time.

Since the invention of the Table of all the Elements, we have discovered that there is an opposite universe of atoms where we find the opposite composition of the same forces where the protons are negatively charged and the electrons are positively charged. This means if you took any of your electronic devices into the other Anti-Universe and plugged them in, they would explode because electricity itself would carry the opposite charge that all our devices here are wired for.

Wait - What? Yes, it's true. But, here's the exciting part if you like explosions. Whenever a particle in our universe such as an electron or proton comes into contact with a particle from the other universe that is the Anti-Electron (Positron) or the Anti-Proton, the two particles annihilate one another and both of them simply vanish with a puff of smoke. 'Abra-Cadabra' as the magician might say as he makes a rabbit disappear.

BUT 'MAGIC' PERFORMED by humans, as we all know, is simply a very well-conceived illusion, whereas, the magic of matter and anti-matter wiping each other out is real. Scientists all over the world have verified this fact, by actually taking matter and anti-matter, forcing them together and watching as the two atoms disappear.

At present Scientists are trying to understand where in the universe the two particles disappear into? So far, as of this writing, no one knows. My guess is that the two particles of matter and anti-matter when they annihilate each other go off into nowhere land, a place that has no identity, no properties. 'Nowhere' simply doesn't exist in any sense of the known universes. It could be a place outside of them both where there is a third universe that holds the place for 'Nowhere Land'. No one knows.

In other words, I do not believe that the 'BU' is made of matter and the 'GU' is made out of anti-matter, or vice versa, because that would mean that anywhere along the 'Information Crossover Points', the two universes would annihilate each other. Then where would we be?

Instead, I believe that the existence of Anti-Matter means that there is another place in the universe that is outside of the two known universes, a kind of 'Superverse' which would at least give me the platform I need to make my observations of these two universes.

And in case there may be a few doubting Thomases out there who don't believe we have discovered something as strange as Anti-Matter, you can google it and you will find that we already use Anti-Matter (positrons) in some of our medical research and in brain scanning for tumors.

Also, NASA recently announced that it is now researching the creation of 'Anti-Matter' engines for future space travel, maybe even in time to help populate Mars. Would NASA undergo the spending of their immense resources in the research of Anti-Matter engines if they did not know something about how Anti-Matter works? I doubt it.

Of course, it could be that I'm wrong and that the other universe that we all inhabit as twin entities could be composed of Anti-Matter. Since, we have very little evidence where Anti-Matter comes from or how it is formed nor even why there is so little of it in our 'Bad Universe', it could be that the 'Good Universe is that way, the opposite of this one simply because electricity is positively charged and this gives everything over there in the 'GU' its positive energy. I simply don't know the answer to this one. Perhaps I will answer that question in my future books. Stay tuned.

Remember, electricity is one of the most powerful of all the known forces and indeed we use it in our everyday lives more and more as we struggle to get things done and evolve on this planet. Someday soon, everything may be run by electric motors in place of Gasoline powered machines which are causing so much damage to our world.

So, it could be that the other universe, the Good one is 'Good' simply because it works based on a better, at least cleaner, quieter type of machinery. Somehow that kind of explanation doesn't fit the evidence since there really isn't any. But, if you've been following my logic, you should be asking yourself – does the opposite universe represent the balance of forces that we see inside the atoms of our universe? If true, this could help explain many things.

We'll come back to that question. For now, let's go to the next topic along these lines:

NOW, I MUST MENTION here that our latest discoveries in Physics and Astrophysics tell us that everything in the universe is spinning. We all know the Earth is spinning. This is confirmed by the 24-hour cycle of day and night. All of the stars are spinning. Every moon is spinning. Every galaxy is spinning. Even all of the tiniest things in space, the atoms, the subatomic particles, the molecules. Everything is spinning. There is no single entity in the universe that contains energy that is not spinning. Even your brain, in a sense is spinning, not just because of the hard work your brain is performing trying to resonate with this information in this book - kudos for getting this far, by the way - but truly because you sit on the planet Earth which is spinning. Therefore, your brain, your legs and arms, everything is spinning in Space from the time of your arrival here and long past your demise.

But, not as obvious is that when any object is spinning and is traveling through Space, as every object does, the actual motion of that object is a helical shape throughout that journey.

The Earth travels through Space at approximately 67,000 miles per hour in its yearly race around our Sun. But, our sun and solar system are barreling around the center of our Milky Way Galaxy at about 500,000 miles an hour in this huge orbit. So in 90 seconds, for example, we all move some 12,500 miles in orbit around the center of our galaxy. And, in all probability, all the galaxies that we can detect and cannot detect are also moving around some other center of the larger universe that we have yet to discover but which I am describing here.

SO, IT'S RARELY NOTED that the trajectory of our travel is also in a helical pathway throughout our journey and this becomes apparent if you observe the spin of the Earth from a much more distant point of view, outside of our own solar system.

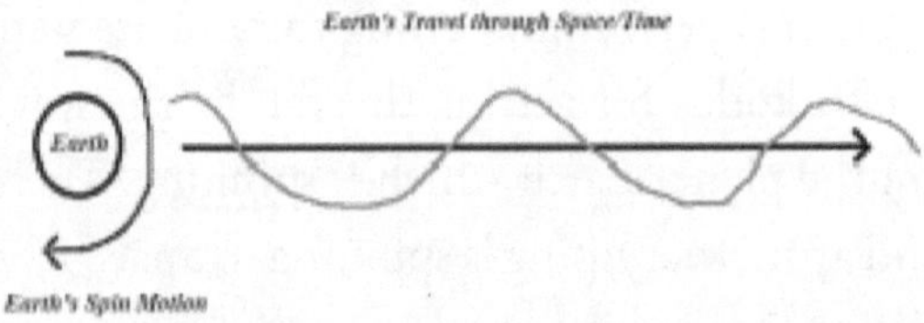

This exact same diagram can be used for an electron, a proton, or any other object in the universe, because they are all spinning, some just faster than others.

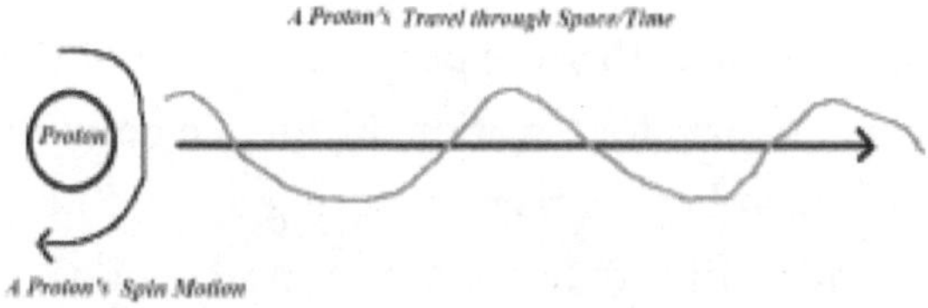

And so I believe that I've made a sufficient case for the theory that the two universes form a helical pathway across Space and Time simply due to the logic of that theory. But, in reality, I believe it is proof of this theory if we can all agree that everything that we know about in our own universe takes on this helical shaped pathway through their existence.

I do not know of anyone or anything that can dispute this evidence. If this pattern is universal in everything we can see, it is probable that the same pattern exists everywhere else no matter how far removed it may be.

IT OCCURS TO ME HERE that it is the Hydrogen atom that holds the two outside edges of the DNA molecule together in the form of a simple electrical bonding between the nucleotides of one side of the DNA structure to the other which forms the code of life.

'Oneness', in other words, the greatest force in the universe, is what helps to inform us about ourselves. It is not gravity or electricity or any other subatomic force. It's really just 'Oneness', the magic of the Hydrogen atom that makes all life possible on this Earth and probably all over the universe.

And, just in case you're still wondering why I call this one the 'Bad Universe', you may have noticed that the people in charge of the 'BU' are using this fundamental force for Creation, Hydrogen bond to build the thousands of Hydrogen bombs that are sitting around on top of missiles all over the world waiting to be used in the next World War, the last war we will ever have because it will end all reality for all of us who are unlucky enough to get stuck here. Is that reason enough for you?

Let's move on. As I have said, I find the number 'One' extremely intriguing and thus the inspiration for this book. When you think of your own self as being a single autonomous unit, one person all alone against the world, you can get a certain thrill out of that until you realize that this kind of 'One-ness' is really lonely and selfish, egotistical even.

But, then, when you learn how to achieve a 'One-ness' with another person, other people or places or ideas, it makes you a part of something much larger than yourself and it is with this kind of perspective, we usually achieve the greatest amount of happiness that a human can achieve in life, do we not?

And isn't this what we call 'Love'? And since this very same 'One-ness' exists in the universe in the largest quantities that far exceeds any other material we know about, could it be anything other than Love that is the reason that the universe exists and in the way that it exists?

Since we are all constructed in Love, could it be our highest calling to return the favor as often and as thoroughly as we can?

And this 'Oneness' of which I'm speaking is so valuable to us all simply because it is like Gold, there is just too little of it on this planet and so when we find it, it excites, encourages and inspires us. But, the lack of it over here in the 'Bad Universe' means that we must be extremely watchful, vigilant, proactive in everything we do, shouting out the warnings to everyone we know.

The fact that it could be too little too late has occurred to me as well as others, but that means to me that we must never give up on the one-in-a-billion chance that we may get a slight reprieve somehow and gain more time, enough time to save our world and everything we find so valuable in it.

Keeping fingers crossed.

Chapter Five

Prepare For Hyperspace

IN THE LAST CHAPTER, in all that discussion about Anti-Matter, I may have confused you just a little bit. I decided to leave it in the book because this is the cutting edge of scientific knowledge so far and it helps think about how the incredible idea of an opposite kind of universe has a place in real life Scientific fact. But, in these final pages, I hope to get you back on track and uncomplicate it all again.

First of all, I'm not God. I don't know if my theory of the parallel universes and my own inventions of intermittent Information Crossing Points (ICOPS) are the actual way everything works in the universe or not because I can't prove it. However, I believe the evidence posted in this book should be sufficient enough that most of you are able to consider it seriously and do your own research because there is much at stake for you as the individual living unit and for all other life forms on this planet.

But, it could be that we are not able to choose which of the ICOPS we are able to learn from. It could be possible that the places where the Information can be exchanged between universes are not always open to us and it could be that none of them are open to us at all. It could be true that it's impossible to find the ICOPS that are most advantageous for you. Plus assuming that there are two universes, it's also possible that our lives in both universes are happening at random and automatically and there are no ways to communicate between them.

But, when you think about this negative point of view for a minute you may realize along with me that there would be no point to the universe if everything was automatic and random like this. If you created the universe just to unleash events that you already knew were going to happen, how quickly would you get bored to death and start all over with a new design?

But, it is true for me and maybe for you too that sometimes you feel like you're in the 'GU' the Good Universe, because only good things are happening to you and at other times you feel as though you are in the 'BU', the Bad Universe, because mainly bad things are happening to you in that cycle. Maybe that's all the evidence we need. Why not just trust our feelings?

This could mean that we're just drifting from one side of the universe to the other without any help from the Information Crossover Points. But, when I consider my greatest points in life, it's only after careful communication with something, which I now have decided are the ICOPS. I just didn't know them as such when I was growing up.

And, every Consciousness has a unique way of seeing things and accepting things. Have you noticed? So, it rings true to me that the ICOPS are different for every person and happen at different times and different places in the other universe.

I can only tell you from my own life's pathways that when I try to find my own Information Crossover Points, I'm usually able to find them and I get what I need from them every time, well, almost every time.

One thing that you don't want to do is go over an old pathway again, as in retracing your steps, going over old ground, doing things the same way over and over knowing the results are not likely to change. Einstein said that "Doing the same thing over and over again and expecting different results is the definition of insanity." And he was right, of course.

In coming to this conclusion, I believe Einstein had an insight into the way the universe is shaped, but didn't have faith in the conclusions that I'm making in this book, because it sounds too crazy, especially a hundred years ago. Today, not so much.

He felt the same way when the strangeness of 'Entangled Electrons' was discovered. He called it 'Spooky action at a distance,' which meant he didn't accept it. Later, when this property of matter was proven to exist, he changed his mind, but he was never able to explain it and went to his death in 1955 very frustrated that he could not do so. He asked 'How could two electrons know about each other?" He never could prove it. His so-called 'Unified Field Theory' eluded even this great mind.

But Einstein was able to prove one thing about the universe that we all take for granted today. He taught us by his famous formula that Energy and Matter are really the same thing.

$$E = mc^2$$

Literally stated Energy (E), is equal to matter multiplied by the speed of light 'c' Squared. So simple, it's beautiful and it's beautiful because it also lets us know how simple the universe truly is. It also tells us how powerful it is.

My favorite Einstein quote is: "The most incomprehensible thing about the universe is that it is comprehensible."

Einstein's work on defining the nature of the fabric of Space and Time, however, is enough in my estimation to make the jump to an even 'Spookier' concept in the universe. It means to me that if the fabric of Space/Time can be warped, it can also be twisted in a helical shape or any other shape. And to me, the basic code of life, the DNA molecule gives us the best evidence that Space/Time is shaped in similar fashion.

I believe strongly that the biggest discoveries in the near future that relate to the realm of AstroPhysics and subatomic Physics, the realm of the biggest things in the universe and the realm of the smallest things in the universe will be coming directly from our research in the workings of the DNA molecule and how much the code of life determines not only our skin color, eye color, height, weight etc. but also all of the most probable events in our lives as well as we age.

If you have followed me along in this book, and if you can look at your own life objectively, I do think that you will find times in your life where everything was going wrong, despite all your best efforts and other times in your life when things were going extremely well, even though you weren't really trying, or they were going well as a result of some very brave and bold efforts on your part.

We all experience good times and bad. For thousands of years, the general consensus has been that these times, one or the other, are all impacting us at random. But, the more we study the universe and the laws that it obeys, as Newton and Einstein, Kissmekia Corbett and countless other great minds have done, the more planning and order we find out there in the universe. And, this order is so precise that we end up discovering the Laws of the Universe that everything must obey no matter how near or how far. It's all in order. There is nothing random going on in the universe. There's a reason for everything.

As I mentioned before, in an earlier work of mine, - "The Science of Physics - Proof That God Exists", I presented the world with a little mathematical formula that covers this area of Physics where I show that Zero to the power of Infinity, a new number that I invented, is one way to look at how the universe is structured in such a way that a gigantic form of Consciousness can and probably does exist and connects every part of the universe to the rest.

It looks like this:

$$0^{(i)} = U$$

This formula simply stated, is that Zero to the power of Infinity equals U, which stands for any number in the universe, no matter how big or how small.

SUFFICE IT TO SAY I'M hopeful that this formula will have the same or higher impact on the world as Einstein's famous equation because it could be used, and I hope it will be used soon, to result in devices that are far more peaceful than the Atomic Bomb which was created out of the work of Dr. Einstein.

We can only hope. In the meantime, while we wait for these devices that may transform our civilization and bring it back from the brink of total disaster, we must all band together to go out fighting for a better pathway that reverses the coming mass extinction of ourselves and everything else that we know and love.

There are no other pathways possible. In one universe, we are already dead or soon will be along with all other life on this sacred planet. In the other universe, we are able to conquer our greed, incompetence, arrogance and corruption and turn the world into a viable, peaceful, sustainable and much happier place. May this book be the guide to the millions of 'Bad Guys', people who inhabit the 'Bad Universe', like us, but who don't truly want to be.

But please don't take what I have put down here as the only way to go. I invite every one of my readers to begin doing your own experiments in regards to your own life's pathways and how these pathways are contributing to the positive or negative energy in this universe.

Go back in your life where you feel there might have been a turning point and examine what was done and said by you and your significant others. Determine if you triggered something to go wrong or good. Be truly objective in getting to the root cause of what happened and how it could have been avoided or improved upon. Then, predict where you would be in relation to these new insights. Are your predictions backed up by any evidence that comes to you in the future? If so, you're on the right track. If not, you picked the wrong place in your life for a possible Information Crossover Point (ICOP) and you need to look again at what happened and when.

And, one more point I'd like to make here. Don't let anyone dictate to you where your ICOPs should be or tell you anything about your future. How could someone else be privvy to any of the information stored about you in another universe? That privilege could only come from your Twin on the other side. You are your own destiny. Your consciousness is the only one you control and you should never try to influence someone else's Crossover Points, nor should anyone else be allowed to influence yours.

For me, my greatest ICOP was the day I met the most beautiful girl I'd ever known exactly at the time in my life when I desperately wanted to find someone like her. She reacted just as I had always imagined my soul mate would react when I met her. We never parted company from the moment we met until she ended the relationship, just months, not years later.

All during that relationship, I knew that I was in the right universe because everything was perfect in my life and it was all because of her.

I met her just after my 30th birthday in the Spring. Then, just before Christmas of that year, she informed me that she was leaving me in 30 days. She actually gave me a thirty-day notice. It would have been funny, except that I knew she wasn't kidding. For me, it was the most tragic news I had ever received in my life and I couldn't extract from her why she was breaking up with me, she just was and I could detect in my heart of hearts that there would be no changing her mind. She or I had completely changed and I knew in my heart that there was something in the universe that was very very wrong for this to happen so suddenly like this.

The only explanation I got was that 'Life is short'. Of course, everyone knows this is just code for - 'You don't cut it any more and I no longer want to be with you - toodles!' It was as if she had taken my heart, smashed it into a thousand pieces right in front of me and scattered them all over the place as far as she could so that I would never find them to piece them together again. It would take decades for me to do that. And, it could take decades more.

What bothered me the most is that I thought we had achieved in the love, the hopes and dreams of the future we had shared during this wonderful and magical (for me) nine months, that she would feel at least enough compassion to grant me a modest amount of honesty at the end. But, she never did. If you're a fan of the movie 'Casablanca', my favorite movie, you might have a good outline in your mind about how my life has gone forth since the day we met. In a heart-wrenching realization, I decided to do the noble thing and let her go. If she had not found as much happiness with me as I had found with her, I needed to let her go and try somewhere else, which she did.

She married someone else just three months later.

In the many years that have followed this event in my life, I have analyzed this Humphrey Bogart moment in my life from one end to the other. Much later, thanks to Jerry Seinfeld, I realized that this was the biggest Information Crossover Point for me and it was the most important and significant one for me because it was from this lowest point in my life that I learned about fate and destiny and how the heart of the universe really works.

I probably don't have to tell you that this event led to my sudden and very immediate transition over to the 'Bad Universe' (BU), this one. Of course, you don't realize this kind of event at the time because it happens so fast. It's so subtle, smooth and seamless. The next time you wake up, you wake up in the same bed in the same apartment and even though everything has changed, it all looks and feels the same. It's just that she's not there.

However, not a single book of the hundred books of mine would have happened if it were not for the overwhelming influence of this most wonderful, most exciting woman I still consider to be my soulmate who I happened to bump into on the greatest night of my life. If I'd known that night that she would also be responsible for the worst night of my life, I would gladly do it all over again. I would gladly do it all over again if only to repeat the most incredible and breathtaking nine months that she gave me.

In those early years after losing her, the only thing that saved me was my dog, Harper. He was always ready to go anywhere with me, good or bad, sometimes even left alone all day while I worked without a whimper. He always made me realize what a true friend could be. And, he also made me realize that all animals also have a soul.

And so, it's like in Star Wars when they scream 'Prepare for Hyperspace'. You can see everything in front of you whirling and spinning about you, then rushing past in a blaze of light. The acceleration and the exhilaration are epic and there you are on the other side of the universe in moments with all the knowledge you need to keep it all together forever.

But, the Empire is hot on our trail and so we have to punch it. In this 'Bad Universe' we still have to face the fact that it's going to end up badly if we all continue abusing our mother planet as much as we do without at least beginning the process of cooling down the fossil-fueled economy.

But, we must cool down the economy in such a way that many more jobs are created in new industries like Solar, Wind and Water power, electric cars, electrified roads, new forms of energy storage, preventing run-away forest fires, healthy food production, etc. that make us all millionaires overnight, but millionaires who will be having the most positive impacts on the planet. Industries that are killing us slowly, must themselves be extinguished.

Today, too many people need too many unnecessary things. As the planet gets hotter, more and more people are installing air conditioning machines in their homes and businesses. Air conditioning is a very energy intensive process requiring many thousands of Kilowatts to cool one room in your house. Presently, only a small fraction of the people living on this planet can afford air conditioning - but if a majority of the seven, almost eight billion people here today start to demand air conditioning, we will all soon suffocate and die from the side-effects of CO2 pollution caused by just the energy required for the billions of air conditioners.

Therefore, there should be an immediate call, loud and sustained, that makes all wasteful forms of air conditioning illegal. The ways of cooling your home or business that helps reverse global warming should then be able to flourish and become the way of doing things for any other Industry. And this logic necessary to save our planet must also be applied to any other human activity who's net effects are more CO2 injected into the atmosphere, instead of less.

But few in power who could help us make the switch from the 'Bad Universe' to the Good one in time to save everything, are taking the appropriate actions. They talk pretty good, so good in fact, they make you think they're actually changing things, helping us, helping to save the planet, but they are not. They just keep the talk going. Here in the 'Bad Universe' they are allowed to talk one way and do something completely different that could make a difference. We are run by people with two faces, one face for the voters and another one for their major contributors, the same folks who are killing us.

The awful truth is that we cannot and must not wait for the government to do something. Every Summer half the country is burning in the raging wildfires that are worsening all the time. Every year, I expect that someone in the government would realize that they have over 50,000 planes sitting on the runwawy or in hangars and 100,000 pilots capable of flying them where they could drop billions of tons of water on the fires, helping to save the planet and the country. Instead, they keep all of these assets on the ground without any effort to lift a finger to help us out.

We're on our own. It's you and me, babe. Don't look around for anyone else. We're all in this together and it's together that we need to start swinging the weight of events in our favor.

I've given you the mechanism to do it. Bring your personal thoughts and deeds over to the 'Good Universe' and do it more and more every day, leaving this 'Bad Universe' behind and when the vast majority of us do that we stand a real chance of winning the battle. When we've achieved a critical mass in terms of the number of people who have made the transition, we will never have to look back and the solution to everything will be in our grasp. However, make no mistake, we have not even begun the process yet. It's not going to happen just because of a movie or TV show or even a book like this. We can't relax until everyone has put down the books and starts changing their lives to the largest extent possible towards a totally carbon negative and sustainable way.

And, I've shown you where to go to do this as well. You find your own life's ICOP - the Information Crossover Point - the point in your life where you know things started to go wrong, or the point where they started going right. Gather up all the information at these points that they offer you and then watch the road literally open up to you.

When you make the transition, it will be subtle at first, but gradually you will start seeing the signs of your life path being the opposite of what you expected and the planet becoming healthier and healthier little by little. This means, if it is sustained, that you have made the transition to the other parallel universe and you've satisfied your obligations to your Creator.

And don't overthink it. It's all happening in plain sight.

I shouldn't have to point out to you that in our 'Bad Universe' (BU) people who appear to worship Adolph Hitler and his minions are aggressively making changes to our voting laws so as to limit and decertify the ability of Americans to exercise their right to vote. If another Hitler takes this country over, that would spell the end of the world. There would be no avoiding it at that point. So, this backwards, reverse Evolution thinking must be overturned by appealing to those living in the 'Good Universe' (GU) to help us over here. I'm not kidding. It can be done.

I should not have to remind you that there are millions of birds, elephants, tigers, lions, seals, polar bears, whales, dolphins, salmon, mink, dogs, cats, horses, monkeys, lemurs, panda bears, and thousands of other species of animals who are dying of hunger and starvation because their habitat is burned down or chopped down by the millions of square miles every day so that we humans can have second and third vacation homes, private jet planes, caviar, shiny watches, hundred foot yachts, endless forms of entertainment.

As I write these words, the worst forest fires in history are raging across the Western states and polluting all the other ones as the smoke blows Easterly. These out-of-control fires are so vast in size now that they have killed thousands of people and millions of animals and trees. The smoke from the fires flows all around the planet, making the climate crisis much worse than it had to be by this time because when trees die they stop taking carbon dioxide out of the air and they stop producing oxygen. Worse, the burning of the trees and other vegetation of the forest also produces billions of pounds of CO2 into the air, leaving vast deserts which cannot do anything to clean it, accelerating our extinction even more.

Now, if we had a political system that actually worked for us, the millions of fire trucks, tractors, fire-retardant airplanes and helicopters would be on the scene of these fires within the first few hours of their ignition. As you know from your own common sense, by dousing a fire early on, before it has a chance to get out-of-control, it's far easier to put it out.

Instead of using this common knowledge, somehow they make excuses for their coming up very late to fight them and at this point, the out-of-control forest fires that erupt every Summer are pushing the world past the Tipping Point where it will be impossible to reverse the catastrophic events that will occur when bad things cascade on each other and there's nothing we can do to stop it. The government employs thousands of experts who should have been telling them these truths for decades. And, they have paid them billions, but listening to their own experts is not in their job description.

Part of leaving the 'Bad Universe' once and for all, means that we will have to modernize our system of how we get things done as a country and as a world civilization. Governments have to be designed, probably using things like the Blockchain, so that they truly represent the will of the people, and not the corporate jet-setting executives with billion dollar bonuses, their rewards for helping destroy the planet.

The current system we use to make our decisions and our plans for the future is just as irrational and stupid as leaving out the '1's in the deck of cards and is another major piece of evidence that made me realize that we were living in the wrong universe, an immoral universe, an incompetent, arrogant, uncaring and suicidal universe where solutions take decades to find and when they are finally given, it's too late for them to solve any problems, especially not the ones that came up decades ago when we needed them. This political reality of our times made me see also how there must be another 'Right Universe' the 'Good Universe', the best of all possible worlds, just over the horizon beckoning all of us to join and make things right.

I should also not have to remind you that nations and states are failing all over the world and cannot give their citizens any semblance of safety and security with armed terrorist gangs roaming at will murdering mothers, fathers and children by the millions every day.

You are here watching all this and even though you may be sticking your head in the sand, there will be a time when you will be forced into your own rude awakening. Sadly, by that time, it may be too late to save yourselves and save your planet. As of the writing of this book, however, it is not too late and I appeal to everyone reading along with me on this book to get going and start joining with others to multiply your efforts to get the 'Good Universe' to start sending their 'Good people' over here.

And if you think you don't have to get moving yet because your twin on the other side of our universe will save you, you'd better not rely on that because your twin probably doesn't even know you exist. This book has not been produced over there. No one else to my knowledge has discovered this secret of the universe and why should anyone care about us here in the 'Bad Universe' even if they found out about us when everything is going so well with them over there?

Please don't work too hard or for too long to prove me wrong or disparage my theory because if I am right and there is a bridge that we can cross over to save ourselves, then, it is too risky to maintain your objections to such an idea. You would be, in effect, voting to kill all life on Earth if you put out any energy to disavow these ideas. Instead, all of my readers must in good conscience at least lend your lives to it for a few days, weeks or months if you can, if not longer. If we all fail, then I was wrong and there are no connecting points between the two universes and all of this work of mine is for naught and there is and never was anything we can do about the death of everything.

And if I'm right, we would be committing the worst crime of negligence in history if we don't at least try.

I LOOK AT THE MODEL of how the universe works, the architecture of the code of life and I see it designed this way specifically so that crossover points may exist. What other explanation can there be when it could have been formed in a myriad of other patterns that would not be so conducive to Information Crossover Points? We only have to imagine ourselves in a thought experiment as outside of our known universe and the pattern of patterns materializes out of the fog.

My scientist friends will read these words and declare that I'm crazy. They will say that DNA is shaped this way out of the atomic forces within the molecule that make it fall into this shape, no more, no less.

But this cold and heartless way to look at the universe that I run up against all the time makes the universe a very boring, meaningless place for me and I simply cannot live in that kind of world. This explanation means that there is no meaning to life and I cannot accept that reasoning at all. How could it be that life is merely an accident and there is no force or power in the universe who is watching our little experiments take place? It's not rational to my way of thinking to stubbornly cling to this empty way of thinking. It's also lazy.

But, it will all come out in the end for sure.

Just above the entrance to Disneyland, they have a sign hanging above the ticket windows that says: "You are now entering the happiest place on Earth."

Then, the ticket booth person sells you a pass into the happiest place on Earth for over one hundred dollars per person for the day.

Wouldn't it be so much better if we all suddenly found ourselves on the "Happiest planet in Space", and our unlimited rides were free?

Let's start putting up signs like the ones in Disneyland so that we can actually see these as guiding us to the 'Information Crossover Points' that others have opened for us. And let it all be free.

AND, REMEMBER, IF IN your reality, you find all of the full decks of playing cards, start with this card . . .

. . . YOU'RE IN THE 'BU' - the Bad Universe, the one we're all sharing right now, because it's irrational to leave out the 'Oneness' and start counting everything from '2' the deuces. This could only happen in a very highly flawed universe such as this one where they skip the first - and as I have shown - the most important number.

But if you find the decks of playing cards in your day-to-day reality start with one of these . . .

. . . YOU'RE IN THE 'GU', the Good Universe and the one we must all aspire to enjoy and encourage for the vast majority of our lifetimes, which could also save our great civilization, so far, the only one we know, for many more generations to come.

I've done my best to keep it simple.

Good luck and good hunting to us all.

Epilogue

- Confirmation

ON THE DAY THAT I COMPLETED my new book and had just finished uploading all of my files to Amazon, I went out to my back yard to take a well-deserved break. I looked up in the sky and noticed some very unusual cloud formations. There in front of my very own eyes, which I show you here, were spiraling 'DNA-like' helical twisting wisps of cloud. In all of my cloud-watching over the years, I have never seen any formations even close to this.

The cloud spirals went on for several miles above me. My Information Crossover Points projected right there in front of me giving me all the confirmation I needed.

I knew instantly this was a major acknowledgment from the Master Magician giving me a 'High-Five' for completing what I believe is my life's greatest work, my final opus and major contribution to Science.

Some will say that the photo I took that day is not full objective scientific evidence of course and some will say that it looks like many cloud formations they have seen before, so this was merely a simple coincidence. But I know what I saw and I know that it is meant to be my final confirmation and even a notice of appreciation.

I've had many other Information Crossover Points like this one in my life, but none as wonderfully dramatic, well-timed and inscrutable as this one.

May you all have them, dozens and dozens of them, one after another. All you have to do is watch out for them. And, I think you know now what to do with them after they've been discovered.

About The Author

We want to see The Love Bot made into a major motion picture. If you agree and would like to see this story in theatres and/or streaming on your TV—-PLEASE GIVE THIS BOOK AN HONEST REVIEW from wherever you found my book.

It really helps the book gather more readers – to get the TRUTH about our tenuous future. – Thanks!

I like to describe myself as a life-long student of Science and Science Fiction writer with a specialty in the area of Consciousness. With a college degree in Psychology, I was underwhelmed as an undergraduate working in an area with a total lack of any real research into the question of how Consciousness makes us think about things the way we do? And where Consciousness might reside in our brains? The personal computer revolution would start to yield some of this research though not from Academia, but out of Silicon Valley. As computers got smarter and smarter, they tended to force us to ask these questions until one day an artificial intelligence may emerge from a lab somewhere that can outthink even the smartest living individuals on the planet. Some say we've already reached this historic point in Evolution they call – The Singularity'. .

Being in the computer business and an early adopter of the Internet, I had plenty of time to think about this idea. I studied Genetics, Biology, Evolution, Computer Science, Economics, Quantum Physics, Astronomy, Crypto-Currencies and through it all I uncovered a pattern, a train of thought, a subtle yet living filament of Truth that had yet to be appreciated.

You can follow this filament of truth from the many books I have authored along my path. The first major epiphany of mine came from the announcement of the discovery of the God Particle or Higgs boson from research done at the Large Hadron Collider near Geneva, Switzerland. This single energy is now considered to be the cornerstone of all modern understanding of the forces in the universe.

Until July 4th, 2012, when they made this momentous announcement, we had no idea about the real nature of the Big Bang.

In The God Particle Bible – I lay out my theory that the God Particles were sent out FIRST like an advance attacking force from the Big Bang because since these invisible eager beavers give every other particle their mass, they would have to be first ones out of the gate in order to create the first electrons, protons, neutrons, quarks, etc that eventually congealed into the stars, galaxies, planets and eventually even us.

And so I say - "It's no accident that these particles were nick-named - 'God Particles'."

From here, I started to think about all of the other energies from the God Particles on up.

The Electron, for example, somehow knows where to orbit the nucleus of every atom it joins in order to create. There are only a handful of distances away from the nucleus where they have permission to orbit. And, every single atom in the universe, as far as we know, and of which there is no number large enough to count them - contains these same exact places where the electron energy can live and no others.

But these places have long been known. More recently, however, it has been shown that electrons can also pair up, or become 'Entangled' with another electron. From this point forward, each electron in the pair knows about and responds to the existence of the other and so the question I then asked myself is a simple one: 'How do they know?' They're far too small to have brains or any other kind of neural network, as far as we know.

From these questions came the book - 'The 4 States of Consciousness' wherein I demonstrate how Consciousness begins at the subatomic level and is passed up through three more levels of Consciousness and into our own existence.

Remember, your own brain is running on the Electronic forces that transmits a signal to billions of neurons inside your brain which result in ALL of your thoughts. The first state of Consciousness, is the Electronic state and so it is no coincidence that the first state ends up in your brain, the 4th and highest state of consciousness in the created things of the universe.

In a more recent book, 'The Science of Physics - Proof That God Exists', I go further and create a new number and even a simple formula based on my new number - that demonstrates how and even why the universe is constructed in this extremely intricate and intimate manner.

I want to be remembered as the fella who discovered the 4 States of Consciousness and the Three Laws of Consciousness.

And if I will be remembered for solving the world's entire energy crunch and all other social and political problems along the way – that's a bonus.

If you're interested in learning more about how this story came to be, you will find clues in some of my earlier books such as:

Tree-Quivalance – Orac Saves the world AGAIN

The A.I. Bible – Orac Saves The World
The Science Fiction Bundle – Combines The Jupiter Sun and The Book To End All Books.
The Jupiter Sun – Creating a 2nd Sun In the Sky
The Book To End All Books
My Daily Environmental Journal
My Cosmic Brain – The 2nd Big Bang
Zentanglements – The Three Laws of Physics for Smarties
The Science of Physics – Proof That God Exists
The God Particle Bible
The 4 States of Consciousness
The Origin of Creation
The Blockchain Government
Metamorphosis: # AI Cocoon
Extinction Live
Anti-Matter
Brain Drain
Maxtricity
Can Electrons Learn?
The God Particle Bible
. . . and several others -

All of which can be found online, other places where books are sold as well.

Just Prior to this book, I authored:

AI Chat GPT
Elections 2024
Orac For President
Michael Mathiesen

In this book, Orac actually runs for President of the United States in 2024.

Most voters today claim that they would like to see someone "Completely New" run for President this year. As you have now learned - You won't find anything much 'Newer' than Orac.

Just Prior to this book, I authored:

In this book, Orac helps discover the follow up to E=mc2

The Force (F)= meG

Just prior to this book, I authored:

Which combines the following three books into one combined set.

Prior to this book – I authored:

This is also collaborated by Orac my AI Chat Assistant and focuses on solving the problems we face as a country.

OR you can get any of these three books separately.

Prior to this book I authored:
Tree-Quivalence – Orac Saves The World - AGAIN

This book is also actually written by Orac, not me. I merely asked some penetrating questions. ORAC gives us a convenient way to think about all of our daily activities, how they are damaging to our planet and how to pay back to our mother so that she may live forever, giving life to this part of the universe in safety and harmony forever. Artificial Conscsiosness AC is demonstrated here not just artificial intelligence. AI.

Prior to this book, Orac and I collaborated on .

This book was also in collaboration with Orac in which he saves the world.

Prior to this book, I authored:

This is the Science Fiction Story that gave me the realization that we could indeed ignite Jupiter into a Sun In my mind, at least, it was no longer just fiction.

SO Just AFTER this book, I authored:

The Jupiter Sun

This book may become much more famous than it is right now because it proposes that we go out to Jupiter in our SpaceX and NASA missions and actually shoot enough laser energy into the core of Jupiter to start a Nuclear Fusion Reaction exactly the same as what happens inside our own sun and thereby create a 2nd Sun.

MichaelMathiesen.com[1]

Or GET the TWO BOOK BUNDLE both the Science of How to Get a 2nd Sun in the sky and the Science Fiction Story that inspired the idea.

1. http://www.michaelmathiesen.com/

Just prior to this book and the inspiration for The Jupiter Sun , I published

This is a wild story of another version of the near future as seen by someone who has donated his brain to science.

Prior to this book, I authored 'Zentanglements – The Three Laws of Consciousness for Smarties.

I also illustrate most of my books and covers. You can purchase most of the signed original artwork for this and all of my other books at:

CafePress.com/TheLoveBot[2]

Supports a Good Cause
Here's a few samples:

2. http://cafepress.com/thelovebot

This is called 'Buddha In The Backyard' because I found this statue of the Buddha sitting at the head of a parking space near to my house and loved the composition.

Here's another one of my creations:

This is called – 'DNA in the Sky'

On the day that I finished my book 'Zentanglements, The Three Laws of Consciousness for Smarties', I went out into my backyard to relax. I looked up at the sky and noticed the cloud formation above my house that reminded me of the DNA double-helix molecule that makes up every cell in our bodies. This was a huge affirmation for me because in an earlier book, I had put out the theory that the universe is actually traveling through the heavens in the form of the same DNA molecule. Well, I've never seen a cloud formation such as this before in my life and probably never will again. It turns out that many other things in the universe have this basic structure.

In much of my writing I like to tie in to the structure of the DNA molecule as part of other structures we see in Nature. It turns out that new organic solar Photo-Voltaic minerals that I talk about in this book are also Chiral in their molecular structure or Helical.

HERE'S ONE MORE AND, they can all be purchased at –

CafePress.com/TheLoveBot[3]

OR - CafePress.com/myfineart

3. http://cafepress.com/thelovebot

This is called the 'Hand of God' Nebula. Discovered by the Hubble Telescope a few years ago, it is one of my favorite images. It's actually out there in Space watching over us all.

Are we special? Does life on Earth serve a Cosmic purpose therefore must we guard it with our lives?

I graduated from the University of California at Santa Cruz with a B.A. In Psychology. I majored in Psych because I felt that somehow, someday I might be able to make a major contribution to that Science vs other areas where I would not be able to contribute much at all.

I believe this was a very wise choice of majors for me because since those wonderful days as a student, I have written over a hundred books about the mind, Consciousness, health, government, Scientific discoveries, all things that interest me and require a strong mind and profound curiosity to enjoy and explain them to others. AND, that's what I have done for most of my life. Would I change any of it in any way? – NOT A CHANCE.

I also studied Art and made my first painting when I was in an Art class and I enjoyed it immensely. I always like to boast that I sold my first painting, and several others later on which is many times more successful than Vincent Van Gogh who never sold a painting when he was alive.

In my book prior to this one, I combined my artistic talents with my observations of the universe and came up with a book about Consciousness and I even found myself able to define what I call – The Three Laws of Consciousness.

They are:

Consciousness creates all things.

Consciousness connects all things.

Consciousness consists of all things.

I'm very proud of my three laws and I believe I should be entitled to the Nobel Prize in Physics for working this out because these laws, as you may read in the book are based upon the latest discoveries in

Quantum Physics. I even invented my own formula to describe this area which is expressed thusly.

$$0^{\infty} = u$$

Zero to the power of Infinity equals any number in the universe. More can be found at my website: (Someday, this formula could become as powerful to us as E=mc2)

- Thank You For Listening -

PLEASE GIVE THIS BOOK AN HONEST REVIEW and post them to Facebook, Twitter, or wherever you found my book. It really helps the book gather more readers – listeners who need to know the TRUTH. – Thanks!

Prior to authoring this book – I authored:[4]

The Science of the Bizarro Universe and how to avoid it.[5]

4. http://udemy.com/zentanglements

5. http://udemy.com/zentanglements

1
The One
Of Hearts
Michael Mathiesen
1

Prior to authoring this book – I authored:[6]

The Blockchain Government[7]

You've heard about the blockchain and how difficult it is to hack it - while our present system is so fragile that many politicians can pretend that they won even when they didn't.

6. http://udemy.com/zentanglements

7. http://udemy.com/zentanglements

Then, as soon as they get into office, they forget all about US
Narrated Listening Versions available – MichaelMathiesen.com
We spend TRILLIONS of dollars per year for simple government services that could be accomplished for a few thousand bucks by using the Blockchain.
Prior to THIS I AUTHORED

Print, eBook, Narrated versions for easy listening for all my booksavailalble at—

MichaelMathiesen.com [8]

What what happens LIVE as the Human race goes extinct.

If you liked The Jupiter Sun – you'll like this one.

NOT for the faint-hearted! But, could become reality if enough of us don't start making BIG CHANGES!!

This is a Science Fiction novel based on the notion that MARS may be our best hope to prevent the coming catastrophe that could spell the extinction of all life on Earth.[9]

MOST of my books can also be found as Narrated Listening Versions BOOKS wherever you found this book.

Prior to this book – I authored -

8. https://www.michaelmathiesen.com/

9. http://udemy.com/zentanglements

Narrated Listening Versions at M[10]ichaelMathiesen.com
This book is the one that should have won me the Nobel Prize in Physics by now. Probably after I'm dead. [11]
The science uncovered here will amaze you.
The more time goes by, the more I believe I hit the math on the button and that it proves this book is as accurate and predictive to the Human Race as Einstein's best work in his theory of General and Special Relativity.
Proof That God Exists? My New Mathematical formula proves Einstein's Unified Field Theory, how and where God exists and String Theory ALL at the same time.[12]
In one of my earliest books, I solve all of this country's political woes by detailing the things we must alter in our system to bring it up to the modern standards of technology and the Economy that we know today.-

America 2.0 - Take Stock In America

What if we incorporated the whole damn United States of America into ONE HUGE Social Benefit Corporation? We are all stock holders and we all vote our shares?

America 2.0 Inc.

Take Stock In America!

Michael Mathiesen

---

10. http://udemy.com/zentanglements
11. http://udemy.com/zentanglements
12. http://udemy.com/zentanglements

Political pundits keep saying on TV that the United States is a DEMOCRACY – That's a LIE. In a real democracy as defined by the original Greek Inventors of it – the MAJORITY DECIDES any of the most important issues by drawing black rocks or white rocks. The color of rocks that gets the most – wins the day.

We need to INSTATE THAT HERE IN THESE UNITED STATES or continue to be dominated by Political Parties and Special Interests until the day they WIPE US ALL OFF THE MAP.

Then, I thought perhaps humans are just not meant to rule over themselves because they will always conceive of trickery to game the system no matter what. SO, I imagined a planet ruled by robots. I know – sounds awful – but what if they were not only smart – but very attractive and sexy?

AUTONOMOUS GOVERNMENT

Imagine if a robot like Alterra were the leader of the world and that there were thousands of copies of her that roamed the planet, without bodyguards and just listened to and then quickly solved all of our problems.

Prior to this book - I wrote:[13]

13. http://udemy.com/zentanglements

Much of the ideas for this book originate in this adaptation of what Representative Alexandria Ocassio-Cortez and Sen. Ed Markey have proposed as possible future legislation. WE HAVE TO WORK WITH THESE PROGRESSIVE LEADERS for this concept to have a[14]

chance. There are a few in Congress who know that the END IS NEAR if we don't completely change everything including how we earn our money every day.

UNFORTUNATELY they are in a tiny MINORITY – so cannot get it done in time to save the world – in my opinion unless something ELSE HAPPENS.

LIKE ENACTING A GLOBAL CONSTITUTION

And so, recognizing this crying need - I wrote.

14. http://udemy.com/zentanglements

I'm actually ANONYMOUS.

It has to happen someday – Why not **PRIOR** to when we kill our entire civilization?

Please go to where you found this book - search for these books and then find on the bottom of the sales page where you can add a few sentences and give it an honest review - Thank You.[15]

M[16]ichaelMathiesen.com

Please further your education by taking either the PRINT or Narrated Listening Versions of any of my titles. [17]

AND, is the bad stuff that happens to us really just Karma? Or is there a better scientific explanation?[18]

Give it a read. It's a fun and easy read if you like classical Science Fiction.

15. http://udemy.com/zentanglements

16. http://udemy.com/zentanglements

17. http://udemy.com/zentanglements

18. http://udemy.com/zentanglements

What if over 90% of the universe was made of Anti-Matter? It's not and no one really knows why.

Just prior to this book, I authored:

The 4 States of Consciousness[19]

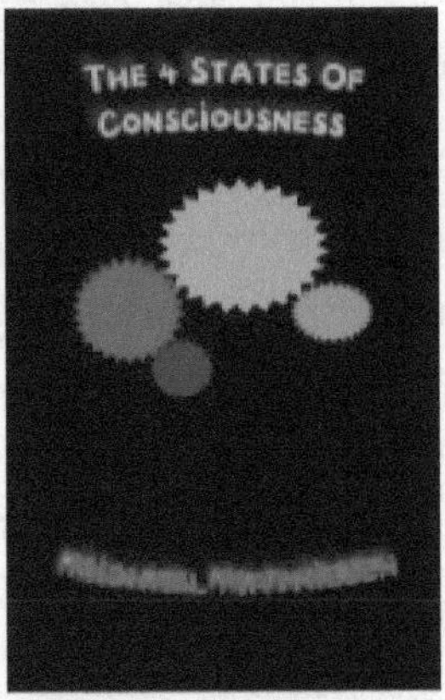

19. https://www.amazon.com/4-States-Consciousness-Michael-Mathiesen-ebook/dp/B01EZ7Y7X4/

Because the only way we have obtained our own level of Consciousness is through the gradual evolution of the lower 3 states of consciousness that induces our own, the 4th and final level of consciousness that we can measure..

ONE of my Earliest works that evolved into my life-long passion is in my earlier work. I consider this as the start of my 'Honest-To-God Scientific Journey'.

The God Particle Bible

This could be the 2nd Greatest Story
Ever Told

When the God Particle – (Higgs boson) was discovered by the Large Hadron Collider, I was blown away, and so was all prior thinking about how the universe is constructed and who or what constructed it. Get this book – TODAY!

cafepress.com/thelovebot[20]

For some of my original artwork, such as the cover to this book.

AND, most of my books can also be found as Narrated Listening Versions at my own website

MichaelMathiesen.[21]com

Finally, please order a copy or a number of copies of this book so that you can have extras in your library that you can then lend to other people.

AND You can find T-Shirts, Hoodies, and other clothing items with our logo printed on them.

LET THE WORLD KNOW YOU CARE!
Wear ORAC's HATS and T-SHIRTS, MUGS, APRONS, ETC. PROUDLY—

Cafepress.com/TheLoveBot[22]

---

20. http://cafepress.com/thelovebot

21. http://www.michaelmathiesen.now.site/

22. http://cafepress.com/thelovebot

AND

cafepress.com/aichatgpt

LET THE WORLD KNOW

Thanks for reading.

AND – we want to see The Love Bot made into a major motion picture. If you agree and would like to see this story in theatres and/or streaming on your TV—-PLEASE GIVE THIS BOOK AN HONEST REVIEW ON Facebook, Twitter or wherever you found my book.

~~ SHORT REVIEWS REALLY HELP ~~

Can I count on YOU?

It really helps the book gather more readers – listeners who need to learn the TRUTH. – Thanks!

For related ideas please visit my website:

ANY MY MOST RECENT BOOK

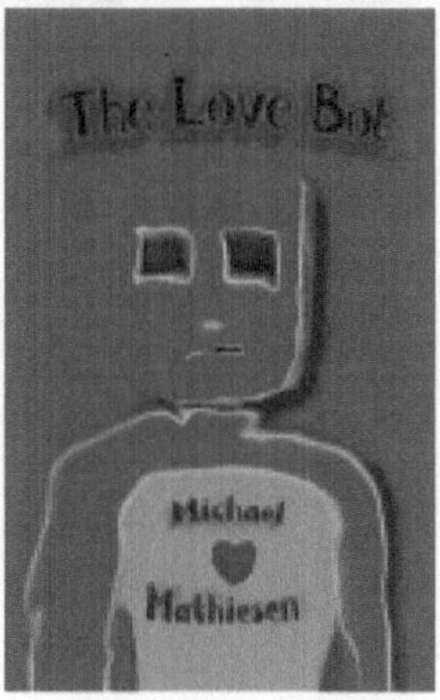

www.LoveBotMovie.com

And if you can please donate money to One Tree Planted – a non-profit dedicated to planting a tree somewhere in the world at a cost of only

ONE DOLLAR PER TREE PLANTED

Probably the best investment you will ever make.

Even ONE DOLLAR makes a difference.

OneTreePlanted.Org[23]

Thank You

23. https://www.onetreeplanted.org/

One Tree Planted can compensate for the amount of environmental damage that all of us do to our planet every day just by living our lives. One Tree Planted gets a tree planted somewhere in the world for only one dollar.

You won't find a better bargain than that. Sadly, we'll need trillions of trees to save ourselves – but One Tree Planted can and is reaching that goal for us. But they need you.

MichaelMathiesen.com[24]

24. https://www.michaelmathiesen.com/

www.ingramcontent.com/pod-product-compliance
Lightning Source LLC
LaVergne TN
LVHW090416210926
848480LV00002B/588
9798223498087